1st International Scientific School on Methods of Digital Image Processing in Optics and Photonics 2014

Journal of Physics: Conference Series Volume 536

St. Petersburg, Russia
8-11 April 2014

ISBN: 978-1-63439-353-9
ISSN: 1742-6588

Printed from e-media with permission by:

Curran Associates, Inc.
57 Morehouse Lane
Red Hook, NY 12571

CURRAN ASSOCIATES INC. proceedings.com

Some format issues inherent in the e-media version may also appear in this print version.

Printed by Curran Associates, Inc. (2014)

Institute of Physics
Dirac House, Temple Back
Bristol BS1 6BE UK

Phone: 44 1 17 929 7481
Fax: 44 1 17 920 0979

techtracking@iop.org

Additional copies of this publication are available from:

Curran Associates, Inc.
57 Morehouse Lane
Red Hook, NY 12571 USA
Phone: 845-758-0400
Fax: 845-758-2634
Email: curran@proceedings.com
Web: www.proceedings.com

TABLE OF CONTENTS

I International Scientific School "Methods of Digital Image Processing in Optics and Photonics" 8–11 April 2014

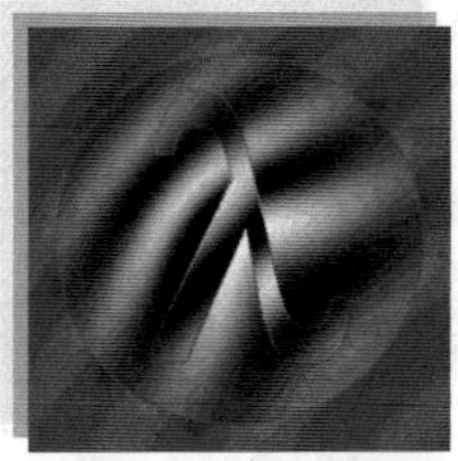

The first international scientific school "Methods of Digital Image Processing in Optics and Photonics" was held with a view to develop cooperation between world-class experts, young scientists, students and post-graduate students, and to exchange information on the current status and directions of research in the field of digital image processing in optics and photonics.

The International Scientific School was managed by:

- Saint Petersburg National Research University of Information Technologies, Mechanics and Optics (ITMO University) – Saint Petersburg (Russia);
- Chernyshevsky Saratov State University – Saratov (Russia);
- National research nuclear University "MEPHI" (NRNU MEPhI) – Moscow (Russia)

The school was held with the participation of the local chapters of Optical Society of America (OSA), the Society of Photo-Optical Instrumentation Engineers (SPIE) and IEEE Photonics Society:

SPIE. CONNECTING MINDS. ADVANCING LIGHT. OSA The Optical Society IEEE Advancing Technology for Humanity

Topics:

- Diffraction imaging
- Interferometry and digital holography
- OCT and 3D vision
- Imaging in scattering and inhomogeneous media
- Hyperspectral imaging
- Super-resolution and sub-diffraction limit imaging
- Optical correlators and pattern recognition
- Optical image and signal processing
- Biomedical imaging

School program:

- Invited lectures from leading Russian and international specialists in the field of methods of digital image processing.
- Reports from young scientists, students and post-graduate students
- Topical seminars and workshop sessions

Journal of Physics: Conference Series **536** (2014) 011001 doi:10.1088/1742-6596/536/1/011001

Invited lectures

M E Gusev, Advantages and new opportunities of digital holographic interferometry, Immanuel Kant Baltic Federal University – Kaliningrad (Russia)
A Kamshilin, Blood pulsation imaging as a tool for studying the cardiovascular system, University of Eastern Finland (Finland)
V Katkovnik, Phase imaging from noisy interferometric observations via sparse coding of phase and magnitude, Tampere University of Technology – Tampere (Finland)
V V Petrov, Construction of the high-resolution acoustic tomography, National Research University of Saratov – Saratov (Russia)
V G Volostnikov, Spiral light beams: properties and applications, Samara Branch of the P.N. Lebedev Physical Institute of the Russian Academy of Sciences – Samara (Russia)

Chairs

Professor I P Gurov (ITMO University)
Professor S A Kozlov (ITMO University)

Scientific committee

O V Andreeva (ITMO University)
V G Bespalov (ITMO University)
P A Cheremkhin (NRNU MEPhI)
K A Grebenyuk (Saratov State University)
I P Gurov (ITMO University)
S A Kozlov (ITMO University)
V V Krasnov (NRNU MEPhI)
N V Petrov (ITMO University)
A S Potapov (Vavilov State Optical Institute)
M A Volynsky (ITMO University)

Scientific Secretary

N V Petrov (ITMO University)

Chairman of the Organizing Committee

M K Khodzitsky (ITMO University)

Organizing Committee

T Yu Nikolaeva, M O Osipova, V A Semenova

Session Chairs

Interferometry and digital holography
N V Petrov, ITMO University – Saint Petersburg (Russia)
K A Grebenyuk Saratov State University – Saratov (Russia)
Broadband, low-coherence methods of image formation and analysis
P A Cheremkhin, NRNU MEPhI – Moscow (Russia)
Khodzitsky M K, ITMO University – Saint Petersburg (Russia)
Representation, encoding, and recognition of images
Krasnov V V, NRNU MEPhI – Moscow (Russia)
Cheremkhin P A, NRNU MEPhI – Moscow (Russia)
Image and optical fields processing algorithms
Volynsky M A, ITMO University – Saint Petersburg (Russia)

The I International Scientific School "Methods of Digital Image Processing in Optics and Photonics" took place at the Department of Photonics and Optical Information Technology of ITMO University, Saint Petersburg, Russia. All sessions were held in the building where Yuri Nikolaevich Denisyuk, the founder of volume holography, worked and conducted research for a long time.

Department of Photonics and Optical Information Technology, ITMO University, Saint Petersburg, Russia

Invited lecture *"Phase imaging from noisy interferometric observations via sparse coding of phase and magnitude"* **V Katkovnik**, Department of Signal Processing, Technology University of Tampere, Finland

Invited lecture *"Benefits and opportunities of digital holographic interferometry"* **Gusev M E**, Immanuel Kant Baltic Federal University, Kaliningrad, Russia

All papers published in this volume of Journal of Physics: Conference Series have been peer reviewed. Reviews were conducted by experts to the professional and scientific standards expected of a proceedings journal published by IOP Publishing. This Proceedings Volume contains 2 invited lectures and 26 reviewed papers, organized by topics and in alphabetical order.

Journal of Physics: Conference Series **536** (2014) 011002 doi:10.1088/1742-6596/536/1/011002

Peer review statement

All papers published in this volume of *Journal of Physics: Conference Series* have been peer reviewed through processes administered by the proceedings Editors. Reviews were conducted by expert referees to the professional and scientific standards expected of a proceedings journal published by IOP Publishing.

Spiral light beams: characteristics and applications

V Volostnikov[1,2], S Kotova[1,2] and S Kishkin[1]
[1] Samara Branch of the P.N.Lebedev Physical Institute, Russian Academy of Sciences, Novo-Sadovaya Str. 221, Samara, 443011, Russia
[2] Samara State Aerospace University, Moskovskoye Shosse 34, Samara 443086, Russia

E-mail: coherent@fian.smr.ru

Abstract. The report contains discussions of theoretical and experimental results of the study on the light beams that retain their intensity structure under propagation and focusing. We describe the behavior of a family of such laser beams called spiral beams, keeping their intensity invariant within scale and rotation, under their propagation. These very features of spiral beams are of a high practical interest for those occupied with the laser technologies and image processing. Some of the applications, in particular their use for the retrieval of the one-dimensional signals phase and contour analysis are demonstrated.

1. Spiral beams

There exists a class of coherent light fields, referred to as spiral beams, of the following form

$$F(x,y,l)=\frac{1}{\sigma}\exp\left(-\frac{x^2+y^2}{\rho^2\sigma}\right)f\left(\frac{x\pm iy}{\rho\sigma}\right), \quad (1)$$

that retain their structure within scale and rotation. Here, $f(z)$ is an arbitrary entire analytic function, $\sigma=1+\frac{2il}{k\rho^2}$ l is the distance along the beam propagation direction, k is the wavenumber, ρ is const, and the sign in the argument of $f(z)$ determines the direction of beam rotation during propagation [1].

This class of the light field contains the beams in the form of arbitrary curves $\zeta(t)$ [2]:

$$\mathsf{S}\left(z,\bar{z}\middle|\zeta(t),\ t\in[0,T]\right)=$$
$$=\exp\left(-\frac{z\bar{z}}{\rho^2}\right)\int_0^T\exp\left[-\frac{\zeta(t)\bar{\zeta}(t)}{\rho^2}+\frac{2z\bar{\zeta}(t)}{\rho^2}+\frac{1}{\rho^2}\int_0^t\left(\bar{\zeta}(\tau)\zeta'(\tau)-\zeta(\tau)\bar{\zeta}'(\tau)\right)d\tau\right]\left|\zeta'(t)\right|dt. \quad (2)$$

The mentioned features are very useful for a number of applications described below.

Journal of Physics: Conference Series **536** (2014) 012001 doi:10.1088/1742-6596/536/1/012001

2. Spiral beams applications

2.1. The analytic continuation for one-dimensional phase retrieval problem [3]

The astigmatic transformation of the light field $U(\xi)exp\left(-\frac{\eta^2}{8\rho^2}\right)$ looks like the following:

$$F(x,y)=\iint_{\mathbb{R}^2}\exp\left(-\frac{ik(x\xi+y\eta)}{f}+\frac{i\xi\eta}{\rho^2}\right)U(\xi)exp\left(-\frac{\eta^2}{8\rho^2}\right)d\xi d\eta = \\ =2\rho\sqrt{2\pi}\exp\left(-\frac{k^2\rho^2(x^2+y^2)}{f^2}\right)\exp\left(-\frac{2ik^2\rho^2xy}{f^2}\right)F_0(x+iy), \tag{3}$$

where the value $F_0(z)=\exp\left(-\frac{k^2\rho^2|z|^2}{f^2}\right)\int_{\mathbb{R}}\exp\left(-\frac{ikz\xi}{f}-\frac{\xi^2}{8\rho^2}\right)U(\xi)d\xi$, f is the focal distance.

It's easy to see that $F_0(z)$ is the analytic continuation of $F_0(x)$ and is determined by means of its zeroes. All the zeroes have the same sign hence it's probable to restore the initial one-dimensional field $U(\xi)$ by a single intensity measurement. In figure 1 the numerical simulation result is presented.

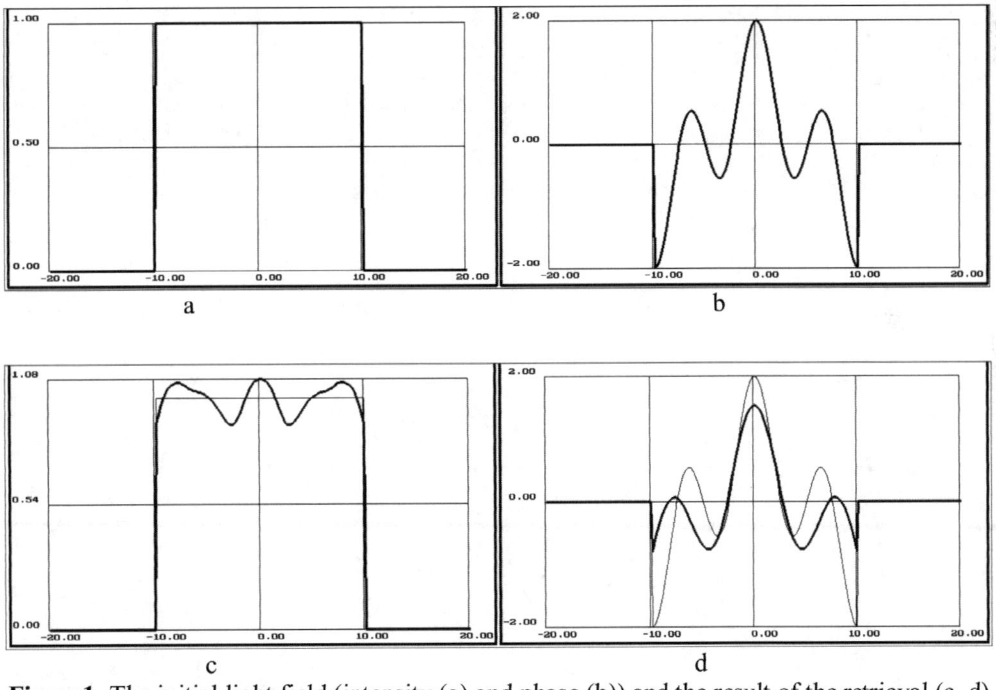

Figure1. The initial light field (intensity (a) and phase (b)) and the result of the retrieval (c, d)

2.2. Contour analysis [4]

In the spiral beams optics a so called quantization condition takes place:

$$S_{curve}=\frac{1}{2}\pi\rho^2N,\ N=0,\ 1,\ 2\ldots, \tag{4}$$

where S_{curve} is the area inside a closed curve. If this condition is valid, then the intensity distribution does not depend on the initial point on the curve. It's obvious, that the same property is typical for any finite sum of the corresponding Taylor series S_N. Note that for the initial contour the dependence of the finite sums on the initial point choice is crucial and represents a serious problem for the methods of the contour image processing.

These circumstances give new possibilities for the contour analysis owing to, the representation of the rotated expansion in the following form:

$$S_N(ze^{i\alpha}, \bar{z}e^{-i\alpha} \mid \zeta(t), t\in[a,a+T]) = \left(e^{-\frac{z\bar{z}}{\rho^2}} \sum_{n=0}^{N} \left(c_n e^{i\alpha n}\right) z^n \right) e^{i\psi(a)} = \left(e^{-\frac{z\bar{z}}{\rho^2}} \sum_{n=0}^{N} c'_n z^n \right) e^{i\psi(a)}, \quad (5)$$

here c_n are coefficients of the Taylor series, $\psi(a)$ is the phase shift.

Thus, the intensity of the rotated beam will be of the same form, and angle of the contour rotation is determined by the following expressions:

$$\forall n \in \overline{0,N}, \ \frac{|c'_n|}{|c_n|} = 1, \ \varphi_n = \frac{1}{in} \ln \frac{c'_n}{c_n} \quad (6)$$

If $\varphi_n = \text{const}$, then this is the angle of contour rotation. Thus the representation of the initial contour as a spiral beam allows us to easily define the angle of rotation irrespective of the initial point of the contour. The outlined method is still being developed now.

References

[1] Volostnikov V G 1997 *Methods for coherent light fields analysis and synthesis: the phase problem study and Gaussian beams optics extension* (Doctoral Thesis, Saratov: Saratov State University)

[2] Abramochkin E G and Volostnikov V G 2004 *Modern Gaussian beams optics* (Moscow: PhysMatLit Publishers)

[3] Volostnikov V G and Loktev M Yu 1999 *Optics and Spectroscopy* **86** 69

[4] Volostnikov V G, Kishkin S A and Kotova S P 2013 *Quantum Electron.* **43** 646

Phase imaging from noisy interferometric observations via sparse coding of phase and amplitude

V. Katkovnik
Department of Signal Processing, Technology University of Tampere, 33720 Tampere, Finland
E-mail: vladimir.katkovnik@tut.fi

Abstract. Phase-shifting interferometry is a coherent optical method that combines high accuracy with high measurement speeds. However, despite its advantageous properties, the inference of the object amplitude and the phase, herein termed wavefront reconstruction is not a trivial task owing to, namely, the Poissonian noise associated with the measurement process and to the 2π phase periodicity of the observation mechanism. In this paper, we formulate the wavefront reconstruction (phase and amplitude) as an inverse problem where the amplitude and the absolute phase are assumed to admit sparse linear representations in suitable sparsifying transforms. We introduce the *sparse phase and amplitude reconstruction* (SPAR) algorithm. SPAR takes into full consideration the Poissonian (photon counting) measurements and uses data adaptive BM3D frames as a sparse representation for amplitude and absolute phase.

1. Introduction

Many images (and signals) admit sparse representations in the sense that they are well approximated by linear combinations of a small number of functions taken from a know set. Fundamentally, this is a consequence of the self-similarity of these images: it is very likely to find in them many similar patches in different locations and at different scales [1].

Let $\mathbf{c} \in \mathbb{R}^n$ denote a vector representing an image, or a patch of it, and let us assume that it admits a sparse representation or sparse coding with respect to the columns of a given matrix $\boldsymbol{\Psi} \in \mathbb{R}^{n \times m}$; i.e., it is possible to write $\mathbf{c} = \boldsymbol{\Psi\theta}$, where $\boldsymbol{\theta}$ is a vector containing only a few non-zero components. The matrix $\boldsymbol{\Psi}$ is termed a synthesis operator (or dictionary) because in the writing $\mathbf{c} = \sum_{j=1}^{m} \boldsymbol{\Psi}_j \cdot \boldsymbol{\theta}_j$, where $\boldsymbol{\Psi}_j$ are the columns of the matrix $\boldsymbol{\Psi}$, and $\boldsymbol{\theta}_j$ are the elements of the vector $\boldsymbol{\theta}$, $\mathbf{c}$ is synthesized as a linear combination of the columns of $\boldsymbol{\Psi}$ weighted by the elements of $\boldsymbol{\theta}$, often called the spectrum of $\mathbf{c}$. The synthesis based representations have a dual point of view in which, given an image $\mathbf{c} \in \mathbb{R}^m$, we compute its spectrum $\boldsymbol{\theta}$ by applying the so-called analysis operator (or dictionary) $\boldsymbol{\Phi}$ to $\mathbf{c}$, i.e. $\boldsymbol{\theta} = \boldsymbol{\Phi}\mathbf{c}$. It happens, that the synthesis dictionaries yielding sparse representations are often overcomplete, i.e., $m > n$. The concept of frame is a generalization of the classical basis especially developed for overcomplete (synthesis and analysis) representations with linearly dependent approximating functions [1]. We extend this idea to complex-valued coherent wave fronts assuming that phase and amplitude of the wave front allow sparse representations.

Let $\mathbf{o} \in \mathbf{C}^n$ be a complex valued wavefront defined on a grid with n pixels. Denote $\mathbf{B}_o =$ mod $(\mathbf{o})$ and $\varphi_o =$angle$(\mathbf{o})$ and as, respectively, the corresponding images of amplitude (modulus) and phase. Then we have $\mathbf{o} = \mathbf{B}_o \exp(j\boldsymbol{\varphi}_o)$. Herein, all functions applied to vectors are to be understood in the component-wise sense; the same applies to multiplication of vectors. With the objective of formulating treatable phase imaging inverse problems, most approaches follow a two-step procedure: in the first step, an estimate of the so-called principal (wrapped, interferometric) phase in the interval $[-\pi, \pi)$ is obtained. The latter procedure is known as phase unwrapping. In what follows we denote the principal phase as $\boldsymbol{\varphi}_o$ and the absolute phase as $\boldsymbol{\varphi}_{o,abs}$ We introduce the phase-wrap operator $\mathcal{W}$ linking the absolute and principal phase as $\boldsymbol{\varphi}_o = \mathcal{W}(\boldsymbol{\varphi}_{o,abs})$, the inverse of this operator denotes the unwrapping, $\boldsymbol{\varphi}_{o,abs} = \mathcal{W}^{-1}(\boldsymbol{\varphi}_o)$.

Following to [2], we introduce the sparse wavefront modeling by the formulas:

$$\text{mod}(\mathbf{o}) = \boldsymbol{\Psi}_{a,o}\boldsymbol{\theta}_{a,o},\ \boldsymbol{\varphi}_{o,abs} = \boldsymbol{\Psi}_{\varphi,o}\boldsymbol{\theta}_{\varphi,o}, \tag{1}$$

$$\boldsymbol{\theta}_{a,o} = \boldsymbol{\Phi}_{a,o}\text{mod}(\mathbf{o}),\ \boldsymbol{\theta}_{\varphi,o} = \boldsymbol{\Phi}_{\varphi,o}\boldsymbol{\varphi}_{o,abs}, \tag{2}$$

where $\boldsymbol{\theta}_{a,o}$ and $\boldsymbol{\theta}_{\varphi,o}$ are, respectively, the amplitude and the phase (absolute phase) spectra of the object $\mathbf{o}$. In (1), the amplitude (mod($\mathbf{o}$)) and phase (angle ($\mathbf{o}$)) are synthesized from the amplitude and phase spectra $\boldsymbol{\theta}_{a,o}$ and $\boldsymbol{\theta}_{\varphi,o}$. On the other hand, the analysis Eqs. (2) give the spectra for amplitude and phase for the wavefront $\mathbf{o}$. Here $\boldsymbol{\Phi}$ and $\boldsymbol{\Psi}$ with the corresponding indexes are the analysis and synthesis matrices (dictionaries), respectively.

The l_o-norm of $\boldsymbol{\theta}_{a,o}$ and $\boldsymbol{\theta}_{a,\varphi}$, denoted as $||\boldsymbol{\theta}_{a,o}||_0$ and $||\boldsymbol{\theta}_{\varphi,o}||_0$, counts the number of non-zero elements of these spectra and is used as the measure of the sparsity of approximation for amplitude and phase, respectively.

2. Optical setup and setting of the problem

We consider the two-beam interferometer shown in fig. 1. A coherent laser beam of the wavelength λ is divided by beam splitter (BS) into two paths, one of which contains a transmitting or reflecting object and the other, a phase shifter. The approach proposed in this paper takes into full consideration the Poissonian (photon counting) measurements. In this way we are targeting at optimal sparse reconstruction both phase and amplitude taking into consideration all details of the observation formation.

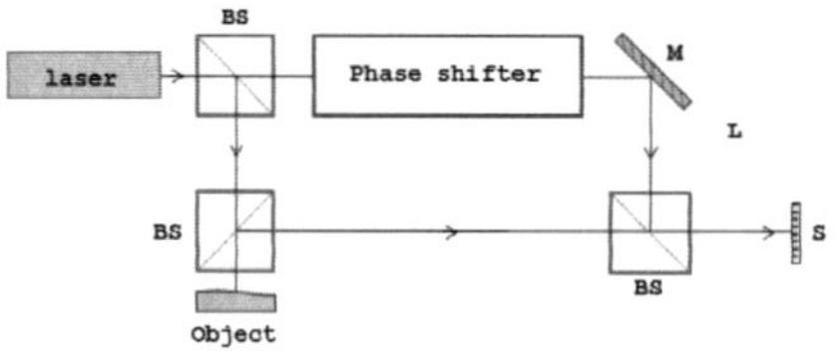

Figure 1. Phase-shift interferometry: on-axis setup. A single laser beam is splitted into two identical beams by a beam splitter (BS). A phase shifter introduces different phase shifts between the object and reference beams.

For the L-step phase-shifting interferometry the complex-valued wave field at the sensor plane is calculated as $\mathbf{u}_s = \mathbf{B}_o \exp(j\boldsymbol{\varphi}_o) + \mathbf{A}_r \exp(-j\varphi_{r_s})$, $s = 1, ..., L$, where $\mathbf{B}_o \exp(j\boldsymbol{\varphi}_o)$ and $\mathbf{A}_r \exp(-j\varphi_r)$ are the object and reference wavefronts, respectively. Let us assume that our sensor takes measures on a rectangular grid with N digital elements and let $\mathbf{Y}_s = \{\mathbf{Y}_s[l],\ l = 1, ..., N\}$, $s = 1, ..., L$, denote the L measured images with N elements each. The measurement process in optics amounts to count the photons hitting the sensor's elements and is well modeled by independent Poisson random variables; that is $p(\mathbf{Y}_s[l] = k) = \exp(-\mathbf{I}_s[l]\chi)\frac{(\mathbf{I}_s[l]\chi)^k}{k!}$, where k is an integer number of photons, $\mathbf{I}_s = |\mathbf{u}_s|^2$ is the intensity of the wavefront, thus given by $\mathbf{I}_s = \mathbf{B}_o^2 + \mathbf{A}_r^2 + 2\mathbf{B}_o\mathbf{A}_r \cos(\boldsymbol{\varphi}_o + \varphi_{r_s})$.

Here χ and is a scaling parameter of the Poissonian flow which can be interpreted as an exposure time (and/or as a sensitivity of the sensor). The problem is to reconstruct the object (wavefront) phase φ_o and the amplitude $\mathbf{B}_o$ from the observations $\{\mathbf{Y}_s,\ s = 1, ..., L\}$. The standard approach to this problem is as follows. The intensities $\mathbf{I}_s$ are replaced by the observations $\mathbf{Y}_s$ and the obtained redundant equations are solved with respect the unknowns φ_o and the amplitude $\mathbf{B}_o$ using the least square method. For the Poissonian observations this procedure is modified because $\mathbf{I}_s$ should be replaced by $\mathbf{Y}_s/\chi$. The scaling parameter χ appears here because $E\{\mathbf{Y}_s\} = \mathbf{I}_s\chi$. In particular, for $L = 4$ and φ_{r_s} taking values $[0, \pi/2, \pi, 3\pi/2]$, phase and amplitude are defined by the equations: $\tan(\tilde{\varphi}_o) = \frac{\mathbf{Y}_4-\mathbf{Y}_2}{\mathbf{Y}_1-\mathbf{Y}_3}$, $\tilde{\mathbf{B}}_o = \sqrt{\sum_{s=1}^{4} \mathbf{Y}_s/4\chi - \mathbf{A}_r^2}$. These estimates $(\tilde{\varphi}_o,\ \tilde{\mathbf{B}}_o)$ usually are quite noisy due to the noisy observations $\mathbf{Y}_s$. The straightforward idea to filter $(\tilde{\varphi}_o,\ \tilde{\mathbf{B}}_o)$ in order to reconstruct $(\varphi_o,\ \mathbf{B}_o)$ is not very productive because the noise in the estimates $(\tilde{\varphi}_o,\ \tilde{\mathbf{B}}_o)$ of phase and amplitude is not additive, and, what is even worse, it is a signal dependent, and this dependence is of complex structure. The iterative algorithm based on the precise modelling of the observations, allows to get precise results even for very noisy data. Comparison of these results versus those obtained by the straightforward filtering of $(\tilde{\varphi}_o,\ \tilde{\mathbf{B}}_o)$ is always in favor of our algorithm.

3. SPAR algorithm

Naturally, the success of the sparse imaging depends on how reach and redundant are the dictionaries $\mathbf{\Phi}$ and $\mathbf{\Psi}$ used for the analysis and synthesis of absolute phase and amplitude. For the analysis and synthesis operations, we use the BM3D frames, where BM3D is an abbreviation for **B**lock-**M**atching and **3D** filtering [2–4]. Herein, we adopt a multiobjective optimization approach to find an object. The main intention of the approach is simultaneous minimization of the minus log-likelihood function of observations (maximum likelihood approach) and the l_0-norms of amplitude and phase spectra supporting the sparsity of the estimates of these variables. The algorithm is derived from three successive stages minimizing three criteria. Overall the algorithm is named **S**parse **P**hase and **A**mplitude **R**econstruction (**SPAR**) algorithm. The details of mathematical formulation of the problems, algorithm implementation as well as the simulation results can be seen in [3, 4].

4. Conclusion

Digital phase-shifting holography is considered. The variational maximum likelihood technique is discussed for phase and amplitude reconstruction of the coherent wavefront from Poissonian intensity measurements. Sparse modeling of amplitude and absolute phase of the wavefront is one of the key elements of the developed algorithm. The experiments show that the new algorithm yields the state-of-the-art accuracy for reconstruction both phase and amplitude of the object wavefront.

Acknowledgments

This work is supported by the Academy of Finland, project no. 138207, 2011-2014.

References

[1] M. Elad 2010 *Sparse and Redundant Representations: from Theory to Applications in Signal and Image Processing* (New York, NY, USA: Springer)

[2] Katkovnik V and Astola J 2012 *J. Opt. Soc. Am. A* **29** 44

[3] Katkovnik V *et al* 2013 Proceedings, Fringe 2013: 7th International Workshop on Advanced Optical Imaging and Metrology, 179

[4] Katkovnik V and Bioucas-Dias J, "Wavefront reconstruction in phase-shifting interferometry via sparse coding of magnitude and phase," J. Opt. Soc., 2014 (in print).

Investigation of phase objects using off-axis digital holography with a-priori known information on the reference wave

A V Belashov[1], **N V Petrov**[2], **I V Semenova**[1], **O S Vasyutinskii**[1,3]
[1] Ioffe Physical Technical Institute, 26, Polytekhnicheskaya, St. Petersburg 194021, Russia
[2] Saint Petersburg National Research University of Information Technologies, Mechanics and Optics, 49 Kronverkskiy pr., St. Petersburg 197101, Russia
[3] St.Petersburg State Polytechnic University; 29, Polytekhnicheskaya, St.Petersburg 195251, Russia

E-mail: belashov.andrey.93@gmail.com

Abstract. Off-axis digital holography with a-priori known information on the reference wave allows phase reconstruction from a single interference pattern, thus, being a promising tool for investigations of dynamic processes. The paper presents the method aimed for estimation of angles between the object and reference wave from a single digital hologram. This information is essential for hologram reconstruction using a-priori known data on the reference wave and may also be beneficial in other reconstruction methods. The technique was validated using a specially prepared phase object; the recording was performed at various angles between the object and reference beams. Main factors affecting the reconstructed phase quality were analyzed.

1. Introduction

Nowadays a wide variety of physical processes is studied using holographic techniques. Holography provides nondestructive recording of spatial and temporal variations of object properties. Analyzing the reconstructed phase one can study particle flows [1], elastic waves in solids [2], thin film vibrations and many other physical phenomena. At present time a wave front reconstruction and phase image processing are performed faster and easier by means of the computer techniques.

Digital photo detector arrays are more sensitive than holographic materials. Besides that, using phase and amplitude digital images one can numerically propagate the wave front backwards and thus get the exact coordinates of all registered amplitude objects. The most common methods of digital holography operate by analogy to that of classical holography: the reconstructed wave front is numerically propagated back to the object plane [3] which results in the formation of diffraction orders containing real and virtual images of the object. Technical characteristics of modern digital sensors used for hologram recording (pixel size and matrix dimensions) impose restrictions on the angle between the object and reference beams and cause reduction of the reconstructed image size. Another group of digital holographic techniques [4] allows one to reconstruct the object wave phase from a set of holograms recorded with a reference wave gaining a particular phase shift from one hologram to another. Such techniques do not provide an instantaneous phase recording and thus are not applicable for studies of fast processes.

2. Off-axis digital holography technique

The method applied in our research is free of these disadvantages and allows one to reconstruct full-size images of the real and virtual parts of the wave front from a single digital hologram [5]. We note however that angles between the object and reference beams are rather small and are difficult to measure. In our experiments a method of angle estimation from the recorded interference pattern had been developed and applied. The algorithm of the object wave amplitude and phase reconstruction is as follows. Assume the reference wave amplitude B, the object wave amplitude A and the phase φ to be constant within a small square area around each pixel. This approximation is feasible for large enough angles between the object and reference beams when the reference wave phase and the interference pattern intensity oscillate sufficiently rapidly. Then the interference of the object and reference waves can be described by the following equation:

$$I = |A\exp(i\varphi) + B\exp(i\alpha)|^2, \tag{1}$$

Applying this equation to each pixel of the square area we obtain an overdetermined system of equations. Unknown parameters can be found then by solving this set of equations using the least-square method, i.e.:

$$\sum_m w_m \, | I_m - | A\exp(i\varphi) + B\exp(i\alpha_m) |^2 |^2, \tag{2}$$

Weight factors w_m adjust an impact of each pixel to the minimized sum; their magnitudes decrease with the distance from the central pixel. That ensures that even if the hypothesis of B, A and φ constancy near the edge of the area under consideration is not fulfilled, it almost does not affect the solution. This sum minimization leads to the system of three linear equations containing three unknown parameters constant within the area [5]. The system solution gives the object wave amplitude and phase at the central pixel. Note that the reference wave phase is supposed to be known. The phase of the object wave cannot be reconstructed without this information.

In most cases it is convenient to use a plane reference wave. So it can be defined precisely from the measured value of an angle between the reference and object waves. Instead of direct measurements of the angle, which is usually difficult due to its smallness, one can simulate the interference of two plane waves and compare the result obtained with the recorded interference pattern. Alternatively the angle can be calculated from a number of interference fringes recorded on the hologram per unit length using the equation:

$$\beta = \arcsin(\lambda / d), \tag{3}$$

where d is the interference fringe width.

Knowing the angle between the object and reference waves one can model the object wave numerically using the equation:

$$B(x) = \exp\left[\frac{2\pi i}{\lambda} x \cos(\beta)\right], \tag{4}$$

3. Experiments and discussion

The described technique was validated in experiments with a specially prepared test object. First the required phase image was modeled on the computer in a form of a grayscale image. This image was projected on the high-resolution photographic negative film. The film was tightly fixed to the dichromated gelatin layer of a holographic plate. Being then exposed by laser radiation from the film side, dichromated gelatin undergoes photolytical dissociation which is stronger in those areas, where the emulsion absorbs more light i.e. where the film transmission is higher. Due to thus induced variations of refractive index in the dichromated gelatin layer, the phase relief copying the grayscale image is formed. To ensure that the test object does not affect the wave front amplitude, intensities of the transmitted laser radiation in several planes were registered, as shown in Figure 1. Since the contour of the test object is visible only in the defocused image (due to light diffraction on the phase relief), the sample indeed does not affect the amplitude component of the light.

The experimental validation of the technique was performed using thus obtained test object. Several digital holograms were recorded with different angles between the object and reference waves. The reconstructed phase images showed good quality for a wide range of angles which indicates that the method applied has less angular restrictions then the Fourier-transformation technique. The major reason of the quality decrease of the reconstructed wave front in our experiments is a low contrast of the fringe pattern. Nevertheless it was shown that even a low-contrast digital hologram can be reconstructed and both the phase information and amplitude image can be obtained.

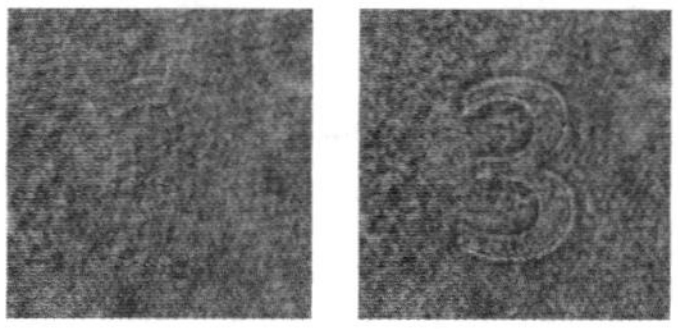

(a) (b)

Figure 1. Transmitted light intensities registered in the test object plane (a) and in an out-of-focus plane (b).

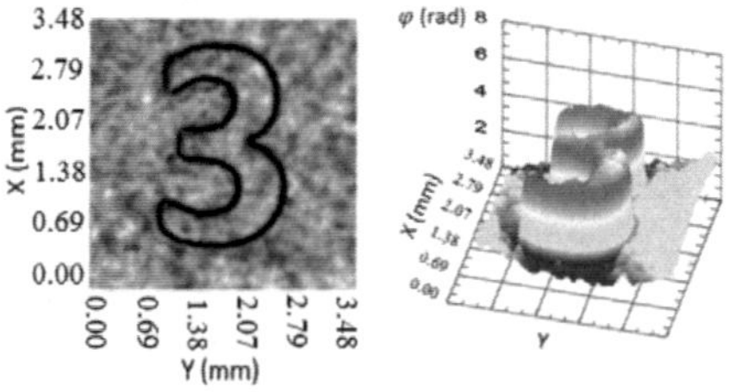

(a) (b)

Figure 2. Reconstructed amplitude (a) and phase (b) of the test object.

Thus, the technique of off-axis digital holography with a-priori known information on the reference wave has been tested with the specially made phase object. The main factors affecting the reconstructed phase quality were analyzed. The method of angles estimation had been suggested allowing to render automatic reconstruction of holograms. The applied technique of complex wave retrieval from a single off-axis hologram possesses several important advantages over more commonly used Fourier transformation and phase shifting holography techniques and, in particular, can provide reliable phase reconstruction in studies of dynamic processes.

Acknowledgments

The financial support from Russian Science Foundation (Project # 14-13-00266) is gratefully acknowledged. Nikolay Petrov acknowledges financial support from Russian Ministry of Education and Science project within the state mission for institutions of higher education (Agreement No 2014/190).

References

[1] Hinsch K D 2002 *Meas. Sci. Technol.* **13** R61.
[2] Dreiden G V, Khusnutdinova K R, Samsonov A M, Semenova I V 2010 *J Appl Phys* **107** 034909
[3] Cuche E, Marquet P, Depeursinge C 2000 *Appl. Opt* **39** 4070-4075.
[4] Yamaguchi I, Zhang T 1997 *Opt. Lett.* **22** 1268-1270.
[5] Liebling M, Blu T, Unser M 2004 *JOSA A* **21** 367-377.

Assessment of transformations of nanoporous volume holograms

R Chigrin[1], **N Andreeva**[1], **O Andreeva**[1]
[1] Saint Petersburg National Research University of Information Technologies, Mechanics and Optics, 49 Kronverkskiy pr., St. Petersburg 197101, Russia

E-mail: rchigreen11@gmail.com

Abstract. The work presents the experimental results on assessment of the impact of the ambient humidity variation on parameters of silver-containing volume holograms in nanoporous silicate matrices 1 mm thick. The experiments were conducted on a dedicated holographic stand with the use of digital methods of data recording and processing. The diffraction angle of 633 nm radiation was found to change by 0.8 mrad at humidity variation in the interval (40÷90%) which is suggested to be related to a change of the average refractive index of a hologram by as much as $6 \cdot 10^{-3}$ and be due to the absorption of water vapors by the light-sensitive composition, distributed in the free pore volume.

1. Introduction

Silicate-glass-based nanoporous matrices are a kind of a special tool for studying physico-chemical processes in a limited volume of a nanosized scale: limited space and effective contact with pore walls determine essential features of filling material state as distinct from the case of its being in a space without any significant restraints.

Porous-matrix-based recording media exhibit a number of useful properties: they are practically shrink-proof and have low coefficient of thermal expansion. Yet, the issues related to changes in the optical parameters of such media have been studied insufficiently.

2. Object of study

The object of study was a transmission volume hologram, recorded in the silver-halide recording medium on the base of a nanoporous silicate matrix about 1 mm thick. The medium exhibits the physico-mechanical properties of silicate glass [1].

The composition of the medium, where the hologram was produced, includes (see Figure 1a): silica skeleton, occupying ~50% of the sample volume; free pore volume ~40% of the sample; gelatin with developed silver particles ~10% of the sample (silver volume concentration ~10^{-4}).

Silica skeleton is known to retain its geometrical and optical properties at variations of the ambient humidity [2]. However, humidity variation can affect the parameters of a hologram, recorded in a silver-halide composition inside the free pore volume. The present work aims to assess the potential changes.

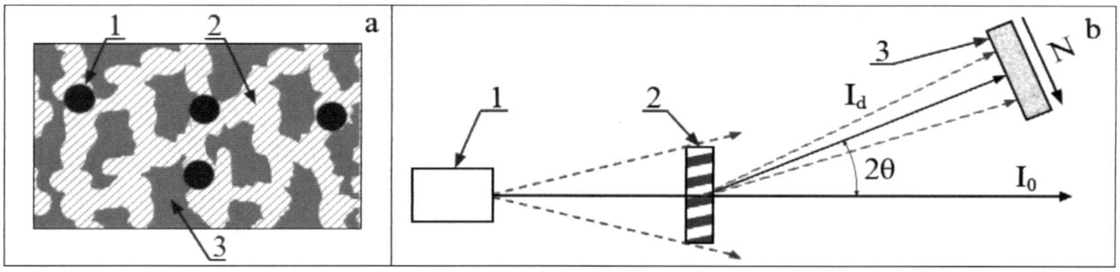

Figure 1. Schematic view of developed silver-halide medium (a): 1 – silica skeleton, 2 – developed silver particle, 3 – immersion, filling free pore volume; (b) - optical schematic of measurements: 1 – radiation source (He-Ne laser, λ=633 nm); 2 – hologram; 3 – CMOS matrix of a photocamera.

3. Experimental procedure

The optical schematic of the experimental setup is given in Figure 1b. The hologram under study was illuminated with a divergent laser radiation beam, and the photocamera matrix recorded the radiation intensity distribution in the diffracted beam (see Figure 2a). Then, the acquired data was digitally processed and the angular selectivity contour of the hologram was constructed (Figure 2b).

Observation and assessment of transformations of the angular selectivity contour of the hologram at variation of the ambient humidity proceeded as follows: formation of the diffracted beam and selection of the frame recording mode (exposure and repetition rate); photocamera recording of necessary number of frames in the automatic mode, computer processing of the acquired data; construction and analysis of angular selectivity contours.

The view of information, recorded in a typical frame, is given in Figure 2a. The angular selectivity contour, derived by processing the frame (by scanning along line A-A) is given in Figure 2c. The transformations were analyzed by comparing the shapes of angular selectivity contours and estimating the change in their positions in the photocamera matrix (see Figure 2c).

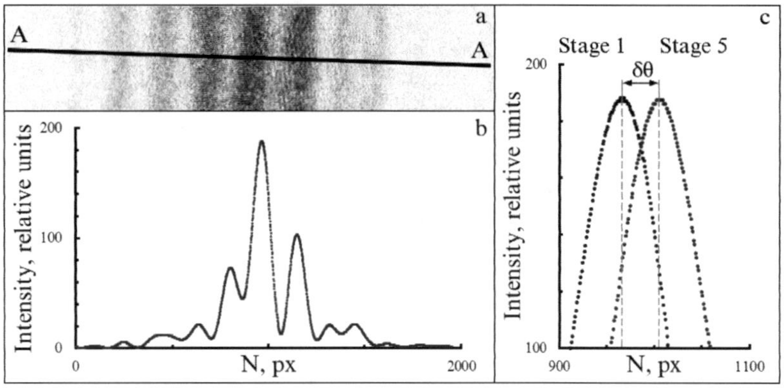

Figure 2. Radiation intensity distribution in the diffracted beam, which was recorded on photocamera matrix (a), experimental data processing results: (b) the angular selectivity contour of the hologram under study, (c) the shift of the angular selectivity contour maximum.

The investigation of the impact of humidity variation comprised several stages (see Figure 3): Stage 1 – installing a sample in the stable state into an optical layout and taking the working frames during several tens of minutes; Stage 2 – creating an enclosure around the sample (with the help of a protective housing), where humidity ~90% was maintained for 20 hours; Stage 3 – taking the working frames at relaxation of a hologram upon the protective housing removal; Stage 4 – exposing the sample under stable conditions in the workroom for 70 hours; Stage 5 – taking the working frames in the stable state of the sample.

4. Experimental results and their analysis

Figure 3 gives the outcomes of processing of experimental results: the position of the principal maximum of the angular selectivity contour of a hologram under study on photocamera matrix is shown for all necessary investigation stages. The quoted data show the recording the frames in the automatic mode with high accuracy to demonstrate the invariability of the position of the angular selectivity contour maximum of a hologram in a stable state (Stages 1 and 5, the second half of Stage 3). Clearly seen is the region of the hologram parameter instability upon a sharp change of humidity – the initial period of Stage 3 of duration about 50 minutes. Attention is drawn to the change in the position of the angular selectivity contour maximum at Stage 5 (stable state of the sample upon exposure in the high humidity atmosphere) against Stage 1 (stable state of the sample before exposure in the high humidity atmosphere). The position of the maximum is clearly seen to have shifted by 40 pixels. The angular resolution of the matrix (the angle between two neighboring pixels) in given layout is 0.02 mrad. Thus, allowing for the experiment geometry, the said shift is an evidence of diffraction angle reduction by 0.8 mrad.

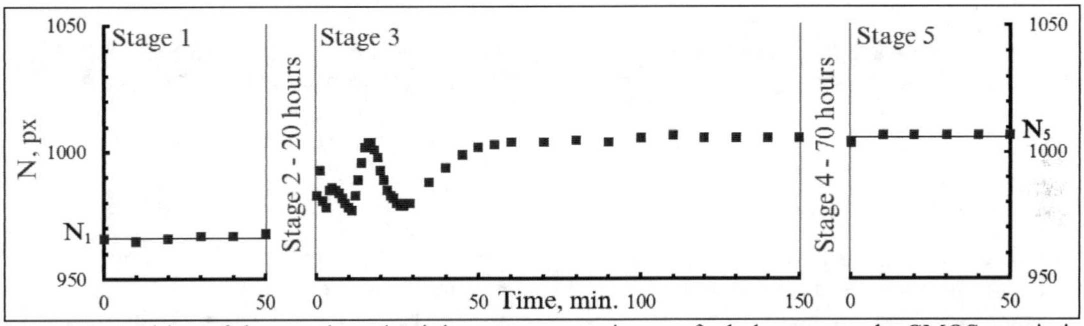

Figure 3. Position of the angular selectivity contour maximum of a hologram on the CMOS matrix in the course of the experiment.

The reconstructing radiation wavelength (laser) and hologram period (due to the silica skeleton) can be considered unchanged during the experiment. Therefore, in conformity with the Bragg condition, the change in the diffraction angle can be attributed to variation of the average refractive index (n_0) of the sample. Under condition of the present experiment, the diffraction angle reduction by 0.8 mrad corresponds to an increase of the average refractive index (n_0) by as much as $5.8 \cdot 10^{-3}$. The detected growth of the average refractive index of the sample upon exposure in the higher humidity atmosphere can be suggested to be attributable to absorption of water vapors in the free pore volume the silver-halide composition occupies.

5. Conclusion

As a result of the present work, the impact of humidity variation on parameters of hologram gratings in nanoporous silicate matrices was assessed. The variation of ambient humidity has been shown to be capable of causing a change in the average refractive index of such a hologram. The analysis of the experimental data allows suggesting the changes in the average refractive index of an air-dry hologram at variation of humidity to be due to absorption of water vapors by the light-sensitive composition in the free pore volume.

References

[1] Andreeva O and Bandyuk O 2011 Light-Sensitive Media-Composites for Recording Volume Holograms Based on Porous Glass and Polymer *Holograms - Recording Materials and Applications* Dr Izabela Naydenova (Ed.) InTech

[2] Andreeva O, Korzinin Yu and Manukhin B 2013 *Holography - Basic Principles and Contemporary Applications* Dr. Emilia Mihaylova (Ed.) InTech

Journal of Physics: Conference Series **536** (2014) 012005 doi:10.1088/1742-6596/536/1/012005

Classical sampling theorem in digital holography

K A Grebenyuk[1]

[1] Department of Physics, Saratov State University, Astrakhanskaya str. 83, Saratov 410012, Russia

E-mail: KonstantinAG@yandex.ru

Abstract. It is proposed to change the present paradigm of the sampling theorem, accepted in papers on digital holography, from 'when recording digital hologram, Nyquist criterion should be fulfilled' to 'when recording on-axis digital hologram, low-pass sampling theorem should be fulfilled, and when recording off-axis one, band-pass sampling theorem should be fulfilled'. Experimental and theoretical basis for this proposal is presented.

1. Present paradigm of the sampling theorem in digital holography

According to Merriam-Webster online dictionary 'paradigm' may mean 'a theory or a group of ideas about how something should be done, made, or thought about'. In the following it will be very convenient to use word 'paradigm' for brief denotation of a group of ideas about how the sampling theorem should be stated, applied and thought about in the field of digital holography.

It is commonly known that when recording digital hologram, requirements of the sampling theorem should be fulfilled. In many papers on digital holography sampling theorem is formulated in one and the same way, as the requirement of having at least two pixels of the digital image sensor per each period of interference fringes (for example, see [1,2]). This requirement can be written in the following form, which we will call 'Nyquist criterion':

$$f_{\max} \leq \frac{f_S}{2} = f_N\ , \qquad (1)$$

where $f_{\max}$ is maximum spatial frequency of the holographic pattern, f_S is sampling frequency of the digital image sensor, f_N is so-called Nyquist frequency of the image sensor or Nyquist limit.

This criterion is usually used without mentioning any conditions of its applicability as something that is inherently valid for the entire digital holography by its nature. Therefore we may say that the essence of present paradigm of the sampling theorem in digital holography reduces to the proposition 'when recording digital hologram, Nyquist criterion should be fulfilled'.

2. Experimental results, which present paradigm does not explain

However there are situations when accuracy of Nyquist criterion at first glance could be doubted. An example of such a case is presented in [3,4].

2.1. Outline of the experiments

In [3] we were experimentally investigating what will happen to reconstructed image if during the process of digital hologram recording spatial frequencies of the holographic structure exceed Nyquist limit. For this purpose we use off-axis lensless Fourier transform digital holography setup.

Numerical reconstruction of lensless Fourier digital hologram with Fresnel algorithm reduces to calculation of its Fourier transform [1]. Thus it turns out that power spectrum of lensless Fourier digital hologram contains reconstructed image of the object.

The object was made as a white paper circle with an "F" letter and was attached to a metal rod with possibility of horizontal lateral displacement along the plane of the image sensor. Shifting the object we were changing its angular displacement θ from the reference point source and thus were changing carrier frequency of the holographic structure, obtained in the sensor plane. Thus we were able to record digital holograms with increasing carrier frequency of the holographic structure.

2.2. Results of the experiments

Figure 1 presents five examples of power spectra of recorded holographic structures. From figures 1(a) and 1(b) we see, that with increase of object displacement θ reconstructed image of the object moves towards the edge of the observed part of the power spectrum, until in the figure 1(c) the carrier frequency of holographic structure reaches the Nyquist limit. After the Nyquist limit is reached, reconstructed image does not vanish, but reappears from the opposite edge of the spectrum (figure 1(d)) and then begins to repeat its previous trajectory throughout the spectrum (figure 1(e)).

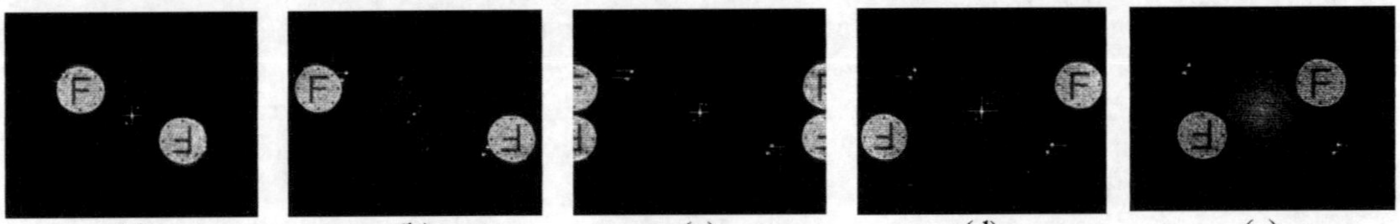

(a) (b) (c) (d) (e)

Figure 1. Power spectra of experimental holographic structures, obtained at different angular displacements θ of the object: (a) θ=1.13°; (b) θ=2.08°; (c) θ=2.71°; (d) θ=3.33°; (e) θ=4.27° [3].

Attempt to interpret these experimental results in frameworks of the paradigm described in the Section 1 leads to contradiction between experimental fact of image reconstruction beyond Nyquist limit and unconditional requirement of the paradigm to observe Nyquist criterion. As a result, the authors of [5], where similar experiments were made, conclude that 'classical Shannon constraints can be relaxed without deteriorating the quality of the image reconstruction'.

3. On the way to new paradigm of the sampling theorem in digital holography

In our opinion clarification of the contradiction described above consists not in recognizing 'incorrectness' or 'excessive severity' of Nyquist criterion, but in recognizing the fact that Nyquist criterion is conditional and simply does not apply to the case under consideration.

3.1. Low-pass sampling theorem

Let's turn to classical works of Kotelnikov [6,7] and Shannon [8] from which commonly used form of the sampling theorem called 'Nyquist criterion' arises. In original work of Kotelnikov [6,7] the sampling theorem is stated as follows (English translation is given according to [9]):

> 'Any function $F(t)$ which consists of the frequencies from 0 to f_1 may be transmitted continuously to any desired degree of precision using numbers which follow on with $1/(2f_1)$ seconds'.

Thus according to its original form sampling theorem is applicable only to signals with frequencies from 0 to f_1, i.e. only to low-pass signals. Following Lüke [9], we will call this original form of the sampling theorem 'low-pass sampling theorem'.

In the process of transformation from original sampling theorem to its commonly used form called 'Nyquist criterion' condition of its applicability has been lost. As a consequence of this loss, it has become a common practice to refer to Nyquist criterion in connection with any setup of digital holograms recording, without taking into account type of signal being recorded.

In experiments with off-axis lensless Fourier holography the spatial signal containing information required for object image reconstruction consisted of the frequencies not from 0 to f_1, but from $f_1 \neq 0$ to f_2, i.e. it was a band-pass signal. That is why low-pass sampling theorem, including Nyquist criterion, is not applicable there.

3.2. Band-pass sampling theorem

Now the question arises: if Nyquist criterion does not apply to sampling of band-pass spatial signals, then what criterion does? The answer is given in [6,7] where in addition to sampling theorem for low-pass signals Kotelnikov also stated sampling theorem for band-pass ones:

'Any function $F(t)$ which consists of the frequencies from f_1 to f_2 may be transmitted continuously to any desired degree of precision using numbers which follow on with $1/[2(f_2 - f_1)]$ seconds'.

For the purpose of distinction let's call sampling theorem for band-pass signals 'band-pass sampling theorem'. The criterion for band-pass signal sampling, corresponding to equation (1), will be

$$\Delta f = (f_2 - f_1) \leq \frac{f_S}{2} = f_N . \qquad (2)$$

It can be seen from the figure 1 that in all five spectra being presented condition (2) is fulfilled, so there is no surprise that image of the object was reconstructed even beyond Nyquist limit. As for gradual decay of the reconstructed image, which occurs with increase of the carrier frequency of holographic structure, it has nothing to do with exceeding the Nyquist limit and is caused only by non-zero value of the pixel fill-factor [4].

4. Summary

Thus from the very origin [6,7] there were two variants of the classical sampling theorem for time signals: one for low-pass signals (consisting of the frequencies from 0 to f_1) and another for band-pass signals (consisting of the frequencies from f_1 to f_2). The same paradigm should be used for space signals sampling in digital holography.

In digital holography the spatial signal, which contains information needed for object image reconstruction is the holographic structure. In the vicinity of each point of the image sensor spatial frequencies of the holographic structure are defined by the angle between the object beam and the reference beam.

In case of on-axis digital holography the holographic structure in general is a low-pass spatial signal and therefore sampling criterion in form of equation (1) should be used. In case of off-axis digital holography the holographic structure in general is a band-pass spatial signal and therefore sampling criterion in form of equation (2) should be applied.

References

[1] Wagner C, Seebacher S, Osten W and Jüptner W 1999 *Appl. Opt.* **38** 4812–20
[2] Kreis T 2002 *Opt. Eng.* **41** 771–8
[3] Grebenyuk K A, Grebenyuk A A and Ryabukho V P 2012 *Proc. of SPIE* 8337 833708-1–5
[4] Grebenyuk K A, Grebenyuk A A and Ryabukho V P 2013 *AIP Conf. Proc.* 1537 102–6
[5] Leclercq M and Picart P 2012 *Opt. Expr.* **20** 18303–12
[6] Kotelnikov V A 1933 *Materials for the 1st All-Union Congress on the Technical Reconstruction of Communication Facilities and Progress in the Low-Currents Industry* (Moscow: Upravlenie Svyazi RKKA)
[7] Kotelnikov V A 2006 *Phys. Usp.* **49** 736–44
[8] Shannon C E 1949 *Proc. IRE* **37** 10–21
[9] Lüke H D 1999 *IEEE Comm. Mag.* 106–8

Modelling natural convection of fluid in cuvette

D Kucher[1], B Manukhin[1], O Andreeva[1], S Chivilikhin[1]
[1]Saint Petersburg National Research University of Information Technologies, Mechanics and Optics, Kronverkskiy pr, 49, St. Petersburg 197101, Russia

E-mail: dmitry.a.kucher@gmail.com

Abstract. Convection is a process of transfer liquid from a hot region to a cool region. This phenomenon is involved in many physical processes. The main characteristic of convection is a temperature field. Modelling of convection allows to get the information about temperature field at any time of process. In this paper the results of modelling natural convection of fluid in cuvette are presented. All results are approved by experimental data. For modelling the process of natural convection Navier-Stokes equations under Boussinesq approximation were used. An experimental setup based on digital holographic interferometry was developedin order to make an experiment. The results for three stadiums of convection, such as: jet initiation, initial jet formation, jet development with formation of mushroom-shaped convective stream, are presented.

1. Introduction

Nanoparticles are used in many advanced areas. Nanoparticles in membrane are the most effective gas separation filters. Semiconductor with nanometre scale shows interesting optical properties. Also nanoparticles are useful in many other fields: medicine, optoelectronics, photonics, etc. Hydrothermal synthesis is one of the methods of formation of nanotubes. Natural convection is one of the important processes in hydrothermal synthesis. Size of the nanoparticles depends on the initial condition of synthesis and changes of physical parameters during the process. Modelling of convection allows to determine the most effective characteristics for synthesis of nanoparticles with predefined sizes. Mathematical simulation of process allows to get the information about temperature, velocity and concentration fields at any time of process [1-3].

2. Mathematical model

Convective motion occurs due to the difference in the density between the heated and cold areas. To describe the process of natural convection it is easy to use Navier-Stokes equations under Boussinesq approximation. We assume fluid density ρ to depend only on the temperature. Navier-Stokes equations and those of continuity and convective thermal conduction in the above approximation are described as follows:

$$\begin{aligned} &\frac{\partial \mathbf{V}}{\partial t} + (\mathbf{V}\nabla)\mathbf{V} = -\frac{1}{\rho_0}\nabla p + \nu\Delta\mathbf{V} + \frac{\rho(T)}{\rho_0}\mathbf{g}, \\ &\nabla\mathbf{V} = 0, \\ &\frac{\partial T}{\partial t} + \mathbf{V}\nabla T = a\Delta T, \end{aligned} \qquad (1)$$

where **V** is the velocity vector, p is the pressure, ν and a are the kinematic viscosity and thermometric conductivity of the liquid, **g** is the free fall acceleration.

The first equation of this system in general case may be rewritten as:

$$\frac{\partial V}{\partial t} = \nu \nabla^2 V - (V \cdot \nabla)V - \frac{1}{\rho_0}\nabla p + f \tag{2}$$

where f is the body force such as gravity or buoyancy.

This equation can be split into two parts (assuming a uniform distribution of pressure):

$$\frac{\partial V_1}{\partial t} = \nu \nabla^2 V + f \tag{3}$$

$$\frac{\partial V_2}{\partial t} = -(V_1 \cdot \nabla)V_2 \tag{4}$$

The first part is diffusion which is described by the equation (3). This equation can be solved by relaxation method. The second part is advection described by the equation (4). It can be solved by MacCormack method, which consists of two steps: the first one is predictor and the second one is corrector.

Using the Helmholtz decomposition, the velocity of liquid can be split into the following form:

$$V = V' + \nabla\varphi \tag{5}$$

where φ is a scalar field and $\nabla V' = 0$.

If we apply the operator∇to both sides of equation (5) and use the second equation of the system (1), we obtain:

$$\nabla^2\varphi = 0 \tag{6}$$

This is a Poisson equation and it can be solved by multigrid method.

3. Results

Implementation of mathematical model allows to get the temperature field as an output data. Real experiment based on digital holographic interferometry was conducted. As a test-object a cuvette with liquid and heater was used. Mathematical description of an object is presented in Figure 1.Boundary conditions for “Insulator” walls are adiabatic. For the top part of object, the boundary condition is "free surface".

In the experimental setup a lamp was used as a heater. Scheme of a real test-object is presented in Figure 2. This scheme includes heater, liquid level and cuvette.

The results of mathematical modeling and experimental data are presented in Table 1. It describes three stages of convection: jet initiation, initial jet formation, jet development with formation of mushroom-shaped convective stream. The time of stages is 3s, 9s, 25s respectively. I – are the interferograms of the object. They were obtained using Fourier hologram scheme. The time shows which images were subtracted to get interferograms.

Journal of Physics: Conference Series **536** (2014) 012006 doi:10.1088/1742-6596/536/1/012006

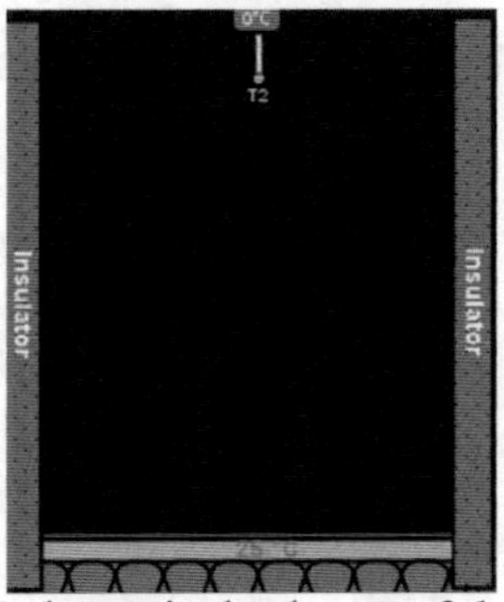

Figure 1. Mathematical scheme of the test-object

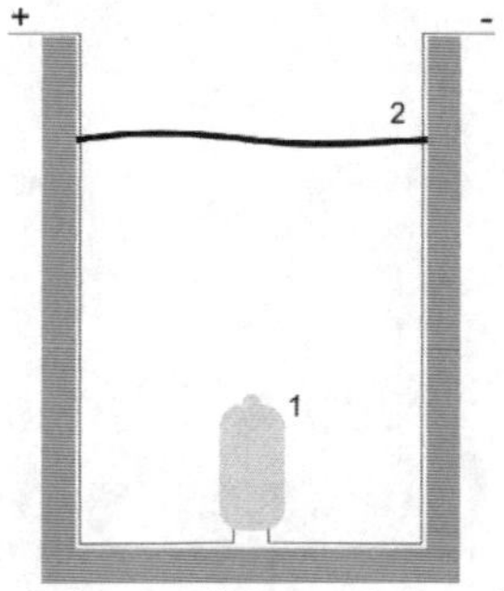

Figure 2. Scheme of the real test-object. 1 – heater, 2 – liquid level

The comparison of the experimental and simulation data for the three stages of convection shows a good result. It allow us to continue our investigation in order to implement the experimental scheme at autoclave.

Table 1. Comparison of the experimental and simulation data. I – interferograms, II – simulation data of the temperature field.

Heating time, s	3	9	25
Stages	1	2	3
I	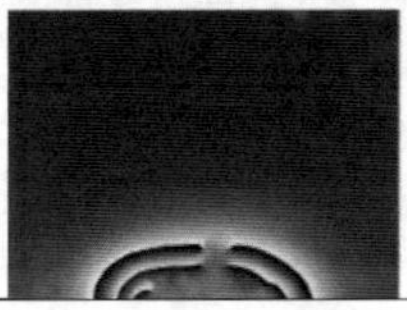	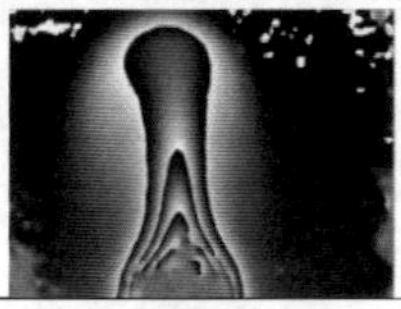	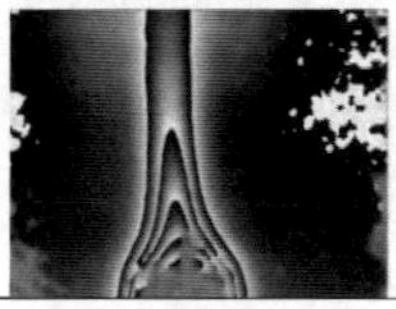
II		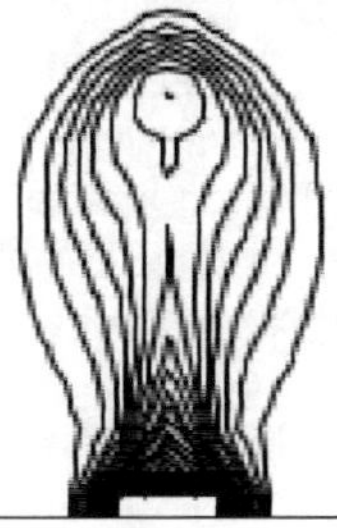	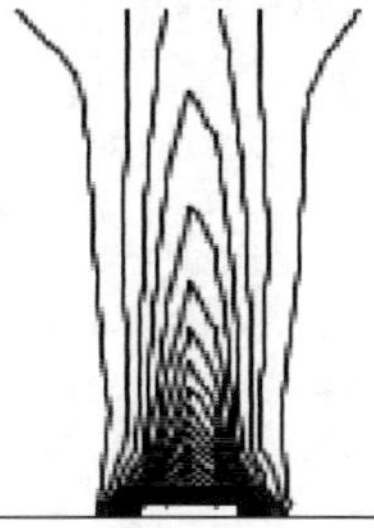

4. Future steps

We consider the implementation of the two-scale model as the next step of our investigation. In our opinion, it will allow to manage micro parameters through macro parameters. Also, weintend to implement particles into our model and add the concentration convection.

References

[1] Aneesh P, Vanaja K, Jayaraj M 2007 *Synthesis of ZnO nanoparticles by hydrothermal method* Proc. of SPIE Vol. 6639

[2] Hiromichi H and Yukiya H 2010 *Hydrothermal Synthesis of Metal Oxide Nanoparticles in Supercritical Water* Materials 2010, 3, 3794-3817

[3] Jianlin Z, Jianglei D, Bingjing W, Jun W, Qian W, Hongzhen J. *Visual and dynamic measurement of temperature fields by use of digital holographic interferometry* / Proc. SPIE 8788, Optical Measurement Systems for Industrial Inspection VIII, 878828 (May 13, 2013)

Study of Thermal Effects in Transparent Objects by Digital Holography Methods

B Manukhin[1], **S Chivilikhin**[1], **G Vinogradova**[1], **M Gusev**[2], **O Andreeva**[1]
[1] Saint Petersburg National Research University of Information Technologies, Mechanics and Optics, 49 Kronverkskiy pr., St. Petersburg 197101, Russia
[2] I. Kant Baltic Federal University, A. Nevskogo st., Kaliningrad 236041, Russia

E-mail: manukhin.ifmo@gmail.com

Abstract. In this article we presented a technique for applying the methods of digital holographic interferometry to analysis the processes, occurring in fluid and solid translucent media under the action o heat sources. The experiments were performed at a small-size stand with the use of a CMOS matrix for recording digital holograms and a computer terminal for data processing. Presented results of investigating the free convection process in a plane water layer and the heat transfer process in a solid, as exemplified by the interaction of radiation with a polymeric recording medium for hologram recording. Features of both processes are examined.

1. Introduction

The study of thermal effects in processes in fluids and solids under the action of external factors is of considerable interest in many scientific and engineering fields [1], as the changes in the temperature of an object lead to those in its optical and physico-mechanical parameters. Investigation of convective fluid flows that accompany most processes in the fluid phase is of great importance in such fields of science and engineering as nanotechnologies, biomedicine, and etc. The study of thermal effects in solids is equally important in research into processes of interaction of radiation with recording media, especially volumetric recording media intended for producing hologram optical elements [2].

2. Experimental setup

In this work we used a stand which is developed for studying transparent objects using methods of digital holographic interferometry (DHI), its optical scheme is given on figure 1. The basis of the development uses the fundamental principles of holographic interferometry [3].

A feature of the experimental stand configuration is setting the unit with the object under study outside the basic interferometry schematic. A considerable elongation of the object beam (as against the reference one) is ensured by great coherence length of the laser radiation used (greater than 10 m).

Placing the object outside the basic optical scheme permits the interferometric studies of the object to be accompanied with other kinds of independent studies (e. g., spectral ones) or the action of external factors on the object to be arranged for.

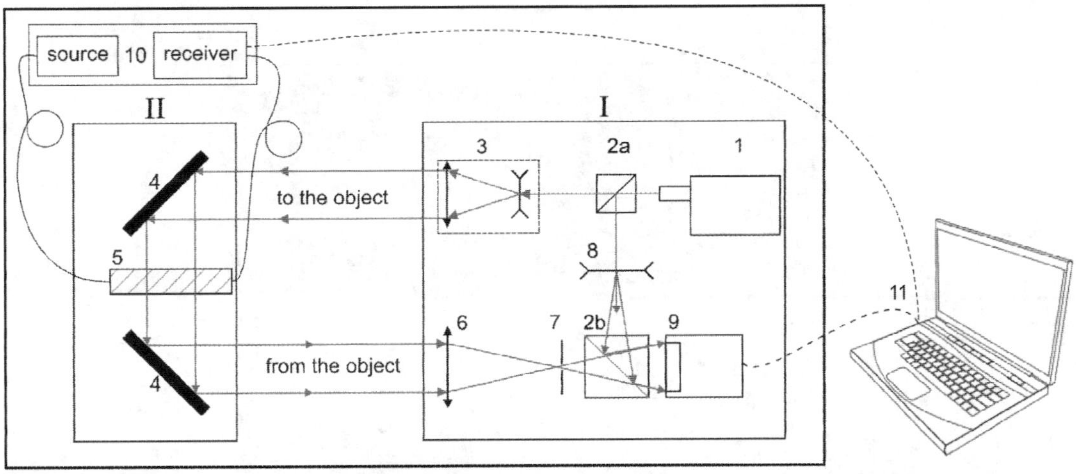

Figure 1. Experimental optical schematic: (1) laser λ = 532 nm, (2) beamsplitters, (3) collimator, (4) mirrors, (5) test object, (6, 8) lenses, (7) diaphragm, (9) videocamera, (10) spectrometer for parallel measurements, (11) computer terminal. I is the interferometer unit, II is the test object unit.

3. Objects of study

To test the developed technique, two processes related to thermal effects were studied: free convection of a fluid under the action of a heart source (A); interaction of radiation with a recording medium, where radiation energy is absorbed and heat propagates from the action region (B).

The object of study in process A is a plane-parallel water layer. Water is an available object suitable as a test object. During the investigation, water was placed in a transparent cell having plane-parallel glass working surfaces, with a heat source inside (tungsten heater in a glass bulb). Wanted convection mode was set by adjusting the heating mode. As the temperature grew, there were no changes in the water layer thickness, the heating was accompanied by variation of the refractive index (in accordance with reference data $dn/dT = -(9.6 \cdot 10^{-5})$ K^{-1}.[4]

The object of study in process B is the samples of polymeric recording medium «Difphen», developed for recording volume holograms [5]. In the course of action of radiation with a wavelength λ = 473 nm upon a sample, energy is absorbed, which is accompanied with local heating of the sample. This results in changes of the average refractive index of the medium and the geometrical dimensions in the action region. Besides, the changes are oppositely directed: linear thermal expansion, $dl/dT = +(3.5\text{-}6.5 \cdot 10^{-5})$ K^{-1}, and refractive index, $dn/dT = -(10.5 \cdot 10^{-5})$ K^{-1}. [6]. Thus, the change of the phase of radiation transmitted through the sample as its temperature varies is due to the difference of coefficients dl/dT и dn/dT and amounts to $d\varphi/dT \approx -(4 \div 7)10^{-5}$ [6].

4. Experimental procedure and analysis of the results

Processing of the holograms includes a number of stages. To begin with, a hologram series is converted to a series of real-time interferograms. To derive each of the latter, the hologram that characterizes the object state at given instant of time is subtracted from the one that characterizes the initial state of the object (before the start of the action). Such a real-time interferogram is used to find the phase difference for given object state and construct a digital interferogram (figure 2, I) by formulas of [3]. Each digital interferogram represents a distribution of phase difference in the object observation plane, which characterizes object state variation. Further computer processing of digital interferogram allows presenting the experimental results and their sampling in the most convenient form.

Figure 2 shows the results of computer processing of experimental data, obtained in the processes under study: digital interferograms (I) and 3D surface of phase difference variation (phase map) (II).

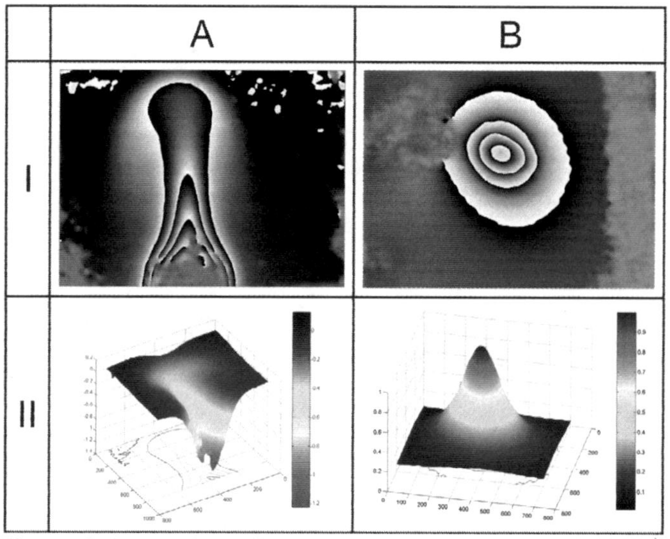

Figure 2. Processing of experimental results: (A) process of free convection in a fluid; (B) process of heat transfer in a solid material. I are digital interferograms, II are 3D surfaces of unwrapped phase difference variation (phase map).

The surface of phase difference variation allows evaluating the temperature field and its changes in the object observation plane. It shall be noted that the calculated surface of phase difference variation (II) indicates the sign of phase variation: with the growing temperature of the sample, the phase advance in process A diminishes, whereas it increases in process B. This disagrees with the values quoted in [6], which were used to estimate the changes in the polymeric base of the sample. Apparently, the PMMA modification used features thermal coefficients differing from tabulated ones, with $dl/dT > dn/dT$ in this case.

5. Conclusion

The presented work implements a technique enabling visual observation and quantitative evaluation of effects due to heat propagation with a source at a local region of the radiation action. The implementation was accomplished on an example of studying the process of free convection in a fluid under the action of a heat source and also a polymeric volume recording medium under the action of radiation, used for hologram recording. The effects of absorbed radiation action were evaluated quantitatively.

The technique can be applied to quantitative evaluation of space-time distribution of the temperature in the object plane within accuracy of tenths of a degree as well as the qualitative assessment of thermal effects on transmissive objects of different parameters.

References

[1] Jianlin Z, Jianglei D 2013 *Proc. SPIE* **8788**, *Optical Measurement Systems for Industrial Inspection VIII*

[2] Andreeva O, Korzinin Yu and Manukhin B 2013 *Holography - Basic Principles and Contemporary Applications, Dr. Emilia Mihaylova (Ed.),* InTech

[3] Pedrini G, Osten W, Gusev M 2006 *Appl. Opt..* **45** 3456-62

[4] Kaye G and Laby T 1986 *Tables of Physical and Chemical Constants* Longman Group

[5] Andreeva O and Bandyuk O 2011 *Holograms - Recording Materials and Applications, Dr Izabela Naydenova (Ed.*), InTech

[6] Weber M 2003 *Handbook of Optical Materials*, CRC Press LLC

Optical reconstruction of digital off-axis Fresnel holograms using phase-only LCOS SLM “HoloEye PLUTO VIS”

P A Cheremkhin[1], N N Evtikhiev[1], V V Krasnov[1], L A Porshneva[1], R S Starikov[1], S N Starikov[1]

[1] National Research Nuclear University MEPhI (Moscow Engineering Physics Institute), Faculty of Experimental and Theoretical Physics, Department of Laser Physics, Kashirskoe shosse 31, Moscow 115409, Russia

E-mail: holo_mephi@mail.ru

Abstract. Optical reconstruction of digital holograms using SLM is used for imaging of 3D scenes, interferometry, microscopy, and etc. In this article reconstruction of digital off-axis Fresnel holograms using phase-only LCOS SLM “HoloEye PLUTO VIS” is described. Experimental and numerically simulated results of reconstruction are presented.

1. Introduction

Spatial light modulators (SLMs) [1] are widely used for displaying of diffractive optical elements. Optical reconstruction of digital holograms is one of the most important tasks [2-5] and is used in different areas of science and industry: imaging of 2D and 3D scenes [2-3], interferometry [4], microscopy [5], and etc.

Since recorded with digital cameras holograms are amplitude type, they have such disadvantages as low diffraction efficiency, necessity of spatial separation of informative diffraction order from undesired orders, and etc. [3]. In addition, quality of reconstructed images from holograms is influenced by noise and linear dynamic range of registering camera [6], characteristics and type (amplitude or phase) of used SLM [7]. Quality of hologram optical reconstruction with phase SLM is worse compared to amplitude SLM. However, application of phase SLM can provide higher diffraction efficiency.

In this article experiments on optical reconstruction of digital off-axis Fresnel holograms displayed on phase-only liquid crystal on silicon spatial light modulator (LCOS SLM) “HoloEye PLUTO VIS” are described.

2. Description of setup for optical reconstruction of digital Fresnel holograms using SLM

Images from digital holograms were optically reconstructed using phase-only LCOS SLM “HoloEye PLUTO VIS” [8]. SLM works “in the reflection” and has following characteristics: quantity of pixels is 1920×1080, pixel size is 8 μm × 8 μm, maximum frame rate is 60 Hz, quantity of phase levels is 256 (8 bit).

Setup scheme for optical reconstruction of digital holograms using SLM is shown in Figure 1. The source of light is a 10 mW HeNe laser with wavelength 633 nm. Laser beam passing through polarizer and collimator consisting of microscope objective, pinhole and collimating lens expands to size of approximately 20 mm in diameter. Then this beam reflecting from displaying digital hologram SLM forms reconstructed wave and passes through analyzer. Computer is used for the SLM control and digital holograms display on the SLM. Scientific digital camera Megaplus II ES11000 was used for reconstructed images registration. Camera is monochromatic and has following characteristics: quantity of pixels is 4008 × 2672, pixel size is 9 μm × 9 μm.

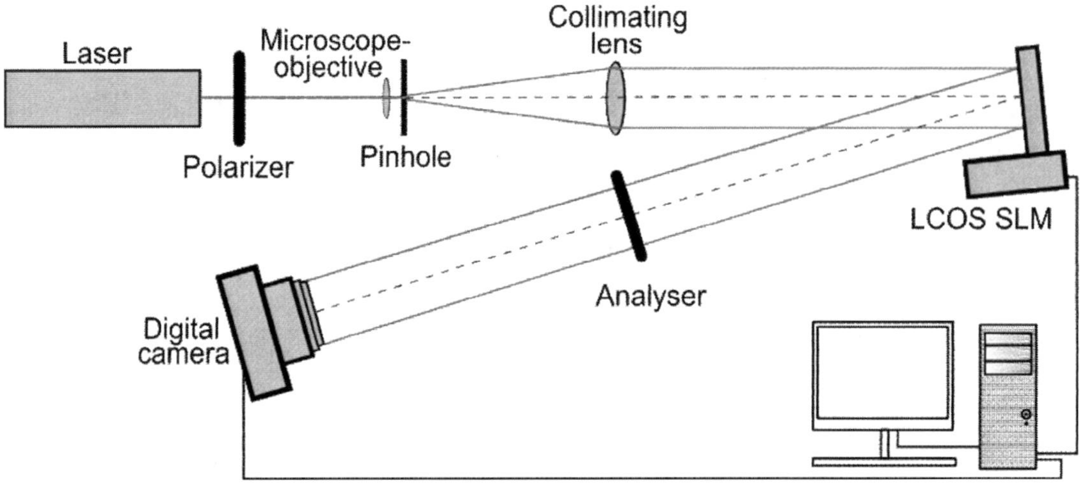

Figure 1. Setup for optical reconstruction of digital Fresnel holograms using LCOS SLM.

Resolutions of displayed digital holograms were limited to the SLM resolution.

3. Experimental and simulated reconstructions of digital Fresnel holograms using phase-only SLM

For experimental and simulated optical reconstruction of digital off-axis Fresnel holograms, earlier recorded [9] holograms were used. Quantity of pixels of these holograms is 2048 × 2048. For experiments, central regions of the holograms with size of 1920 × 1080 pixels were chosen. Holograms were optically reconstructed with its display on the phase-only SLM. Fragments of reconstructed images were registered by camera Megaplus II ES11000. Example of reconstruction of one hologram is shown in Figure 2a. Higher noise level on left part of image is caused by presence of undesired zero diffraction order.

For comparison, Figure 2b shows image fragment obtained by modeling of optical reconstruction of hologram using phase-only SLM. Modeling of optical reconstruction was carried out as follows. Firstly quantity of pixel signal levels was decreased to 256. Then modeling of hologram displaying on phase-only SLM was performed. It was supposed that value of phase shift introduced by pixel of SLM is directly proportional to signal value of pixel of digital hologram. As a result, amplitude of modeled wave that incident onto SLM did not change after transmission through the SLM or reflection from SLM. Only its phase changed. Modeling of wave propagation from SLM plane to plane of image reconstruction was accomplished with method of direct calculation of Fresnel diffraction (through one fast Fourier transform [3, 9]).

As seen from Figures 2a and 2b, quality of experimental reconstruction is quite similar to results of numerical modeling. Higher level of noise of experimentally reconstructed image is due to temporal fluctuations of phase modulation of LCOS SLM [10].

a b
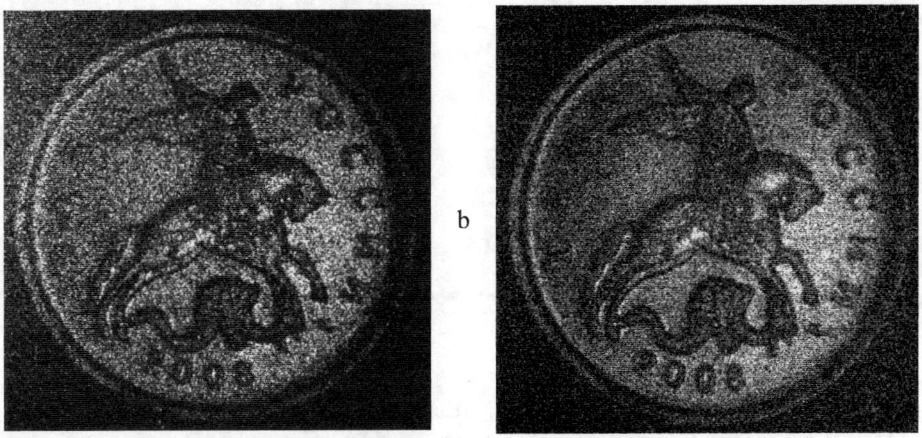

Figure 2. Experimental (a) and numerically simulated (b) digital Fresnel hologram reconstruction using phase-only SLM.

To decrease temporal fluctuations of SLM, synchronization of SLM and registering camera can be used [11]. This will allow to improve quality of digital hologram reconstruction.

4. Conclusion

Experiments on optical reconstruction of digital off-axis Fresnel holograms using LCOS SLM were performed. Experimental and simulated results of reconstruction are compared and are close to each other. Synchronization of SLM and registering camera is proposed to improve quality of experimental reconstruction.

Acknowledgments

This work was partially supported by the Russian Foundation for Basic Research (RFBR) № 13-07-00395.

References

[1] Wu S T and Yang D K 2005 *Reflective Liquid Crystal Displays* (Chichester: John Wiley)
[2] Zwick S, Haist T, Warber M and Osten W 2010 *Appl. Opt.* **49** F47-58
[3] Juptner W and Schnars U 2005 *Digital Holography: Digital Hologram Recording, Numerical Reconstruction, and Related Techniques* (Berlin-Heidelberg: Springer-Verlag)
[4] Demoli N 2006 *Opt. Express* **14** 2117-22
[5] Bouamamaa L, Bouafiaa M, Wernicke G, Krueger S and Gruberet H 2004 *Catal. Today* **89** 337–41
[6] Evtikhiev N N, Starikov S N and Cheremkhin P A 2013 *J. Opt. Technol.* **80** 301-8
[7] Evtikhiev N N, Starikov S N, Cheremkhin P A and Kurbatova E A 2014 Estimation of diffraction efficiency and image quality of optical reconstruction of digital Fresnel holograms *Radiophys. and Quantum Electronics* **56-57** (to be printed)
[8] http://www.holoeye.com/spatial_light_modulators_pluto.html.
[9] Evtikhiev N N, Starikov S N, Cheryomkhin P A, Krasnov V V and Rodin V G 2012 *Proc. SPIE, Optical Modelling and Design II (Brussels)* vol 8429 (SPIE) p 84291M
[10] Lizana A, Moreno I, Márquez A, Iemmi C, Fernández E, Campos J and Yzuel M J 2008 *Opt. Express* **16** 16711-22
[11] Cheremkhin P A, Evtikhiev N N, Krasnov V V, Starikov S N and Rodin V G 2014 *Proc. SPIE, Practical Holography XXVIII: Materials and Applications (San Francisco)* vol 9006 (SPIE) p 900615

A Comparison of Terahertz Pulsed Spectroscopy and Backward-Wave Oscillator Spectroscopy

Kirill I. Zaytsev[1], Arseniy A. Gavdush[1], Nikita V. Chernomyrdin[1], Nikita P. Kruchkov[1], Konstantin G. Kudrin[1,2], Pavel A. Nosov[1], Valeriy E. Karasik[1], and Stanislav O. Yurchenko[1]

[1]Bauman Moscow State Technical University, 105005, 2nd Baumanskaya str., 5/1, Moscow, Russia

[2]P.A.Hertzen Moscow Research Institute of Oncology, 125284, 2nd Botkinsky proezd 3, Moscow, Russia

E-mail: kirzay@gmail.ru

Abstract. This paper presents new experimental and theoretical results for the material parameter reconstruction utilizing the terahertz (THz) pulsed spectroscopy (TPS). The material parameter reconstruction algorithm was realized and experimentally implemented to study the test sample. The algorithm takes into account multiple reflections of THz pulse within the flat sample during the transmission mode measurements. Therefore the samples with small thickness or low refractive index could be studied utilizing the proposed method. In order to estimate the reconstruction accuracy, test sample material parameters, obtained with the TPS, were compared with the results of the same sample studying by the use of the backward-wave oscillator (BWO) spectroscopy. Thus, high reconstruction accuracy was demonstrated.

1. Introduction

Terahertz (THz) pulsed spectroscopy (TPS) is proved to be a convenient instrument for characterizing materials in THz frequency range [1]. TPS utilizes short pulses of THz radiation with a wideband spectrum in frequency-domain to measure the THz spectral material parameters [2–4] or to study the internal structure of the sample [5]. TPS signal processing and TPS inverse ill-posed problem solution are important for various fields of physics. In TPS, the electric field of the THz pulse is detected with a high time resolution after the transmission of a pulse through the sample or after the reflection of a pulse from the surface. By applying a fast Fourier transform to the measured THz waveforms, it is possible to analyze the complex transmission or reflection coefficients of the sample in a wide frequency range. Depending on the techniques, utilized for the THz pulse generation and detection, the reliable frequency range of TPS operation could cover different ranges of the electromagnetic spectrum.

There are many applications for THz technology and, in particular, TPS systems. The list of applications includes spectroscopy of semiconductors [6], gas sensing [7], security task [8–10], and medical diagnosis (diagnosis of skin cancers [11–13], skin burns [14, 15], breast tumors [16], cancer of colonic tissue [17], etc.). Recently, TPS has become a useful tool for the nondestructive evaluation of polymer constructional materials and composite structures [18, 19].

In this paper, the new experimental and theoretical results for the material parameter reconstruction using TPS system are presented and discussed. A common material parameter reconstruction technique [2–4] is realized and experimentally implemented for studying the test sample optical properties. In order to both verify the material parameter reconstruction procedure and to estimate the reconstruction accuracy, obtained material parameters of the test samples are compared with the results of the same samples studying by the use of the backward-wave oscillator (BWO) spectroscopy [20–23].

2. Materials and Methods

The conventional TPS system [5, 24], which is based on the generation of THz pulses in PC antenna and detection of THz waveforms using electro-optical sampling, is used. The TPS system works in transmission mode allowing determination of the sample complex transmission coefficient in the frequency range between 0.1 and 3.0 THz.

A common material parameter reconstruction procedure, which is similar to the one described in papers [2–4], is used as the basis for the material parameter reconstruction. This technique helps to determine the spectral complex refractive index $\widetilde{n}_1(\nu_t)$ of the flat sample on the basis of the experimental sample complex transfer function $\widetilde{H}_{exp}(\nu_t)$. Note, multiple reflections of THz radiation within the flat sample, leading to the appearance of the satellite THz pulse in TPS waveforms, are considered [4]. Thus, flat samples with small thickness and low refractive index could be studied.

The sample thickness l needs to be measured before the TPS waveform processing with an accuracy better than 0.05 mm. Considering the typical measurement conditions, the flat samples, which are transparent in the THz frequency range and have a thickness greater than 100...200 μm, could be studied. The list of important conditions for correct algorithm implementation also includes height sample flatness, sample homogeneity, and small roughness of the sample surface. The maximal angle deviation between the flat surface normal and the axes of the THz beam should be smaller than 5° [2].

Material parameter reconstruction is produced by means of the error functional $\Phi(\nu_t, \widetilde{n}_1(\nu_t))$ minimization:

$$\widetilde{n}_1(\nu_t) = \arg\min_{\widetilde{n}_1(\nu_t)} \Big[\Phi\big(\nu_t, \widetilde{n}_1(\nu_t)\big)\Big] = \arg\min_{\widetilde{n}_1(\nu_t)} \Big[M\big(\nu_t, \widetilde{n}_1(\nu_t)\big) + A\big(\nu_t, \widetilde{n}_1(\nu_t)\big)\Big], \qquad (1)$$

where $M(\nu_t, \widetilde{n}(\nu_t))$ and $A(\nu_t, \widetilde{n}(\nu_t))$ depend on the experimental $\widetilde{H}_{exp}(\nu_t)$ and theoretical $\widetilde{H}_{th}(\nu_t, \widetilde{n}(\nu_t))$ transfer functions of the flat sample

$$\begin{aligned} M(\nu_t, \widetilde{n}_1(\nu_t)) &= \left| \left|\widetilde{H}_{exp}(\nu_t)\right| - \left|\widetilde{H}_{th}(\nu_t, \widetilde{n}_1(\nu_t))\right| \right|^2, \\ A(\nu_t, \widetilde{n}_1(\nu_t)) &= \left| \phi\left[\widetilde{H}_{exp}(\nu_t)\right] - \phi\left[\widetilde{H}_{th}(\nu_t, \widetilde{n}_1(\nu_t))\right] \right|^2. \end{aligned} \qquad (2)$$

where $|...|$ and $\phi[...]$ represent the modulus and the phase of the complex functions, respectively.

Functional minimization (Eq. (1)) could be implemented separately for different frequencies ν_t of the discrete Fourier-domain. Beginning at a certain ν_t', which corresponds to the highest TPS sensitivity, we can calculate the sample complex refractive index $\widetilde{n}_1(\nu_t')$ and use it as an initial assumption for the neighboring $\widetilde{n}_1(\nu_t' + \Delta\nu_t)$ computation, where $\Delta\nu_t$ is a step of the discrete Fourier-domain. This approach for $\Phi(\nu_t, \widetilde{n}_1(\nu_t))$ minimization excludes an appearance of non-physical breaks in reconstructed material parameter curves.

The experimental complex transfer function $\widetilde{H}_{exp}(\nu_t)$ can be measured using the relation

$$\widetilde{H}_{exp}(\nu_t) = \frac{\widetilde{E}_s(\nu_t)}{E_r(\nu_t)}, \qquad (3)$$

where tildes stand for Fourier transformation: $\widetilde{E}_{r,s}(\nu_t) = F_t^{-1}[E_{r,s}(t)]$, and the reference $E_r(t)$ and sample $E_s(t)$ waveforms are introduced.

Theoretical transfer function is based on the quasi-Fabry-Perot model [2–4] of the wave transfer through the sample:

$$\widetilde{H}_{th}(\nu_t, \widetilde{n}_1(\nu_t), l, N) = \widetilde{P}_0(\nu_t, \widetilde{n}_1(\nu_t), -l)\, \widetilde{T}_{01}(\nu_t, \widetilde{n}_1(\nu_t)) \times$$
$$\times\, \widetilde{P}_1(\nu_t, \widetilde{n}_1(\nu_t), l)\, \widetilde{T}_{10}(\nu_t, \widetilde{n}_1(\nu_t)) \left[1 + \sum_{\alpha=1}^{N} \left[\widetilde{P}_1^2(\nu_t, \widetilde{n}_1(\nu_t), l)\, \widetilde{R}_{10}^2(\nu_t, \widetilde{n}_1(\nu_t))\right]^{\alpha}\right], \tag{4}$$

where N represents the number of multiple THz pulse reflection within the flat (number of satellite pulses in sample waveform $E_s(t)$). In (4), the complex operator $\widetilde{P}_\beta(z, \nu_t)$ describing the wave propagation through the medium along the distance z and defined by the Bouguer-Lambert law is

$$\widetilde{P}_\beta(z, \nu_t) = exp(-i2\pi\widetilde{n}_\beta(\nu_t)z), \tag{5}$$

$\widetilde{T}_{ij}$ and $\widetilde{R}_{ij}$ in (4) are a complex operators describing the transmission and the reflection of the THz pulse at the interface between the i-th and j-th media and defined by the Fresnel relations

$$\widetilde{T}_{ij}(\nu_t) = \frac{2\,\widetilde{n}_i(\nu_t)}{\widetilde{n}_i(\nu_t) + \widetilde{n}_j(\nu_t)}, \qquad \widetilde{R}_{ij}(\nu_t) = \frac{\widetilde{n}_j(\nu_t) - \widetilde{n}_i(\nu_t)}{\widetilde{n}_i(\nu_t) + \widetilde{n}_j(\nu_t)}. \tag{6}$$

The sample waveform contains a limited number of satellite pulses N owing to the finiteness of the TPS time-domain window; therefore, a finite number of pulses should be taken into account during the $\widetilde{n}_1(\nu_t)$ computations. An effective sample thickness correction, based on criterion of the material parameter curve smoothness [2, 4], is applied during the material parameter reconstruction to increase the measurement accuracy.

3. Material Parameter Reconstructed with TPS and BWO spectroscopy

BWO spectroscopy is a very precise tool for determining material parameters at the GHz and THz frequencies. It has attracted an extensive amount of research on the methodology of dielectric measurements with BWO spectroscopy [21–23, 25]. However, BWO spectroscopy suffers from a limited sensitivity range ($\nu_t \leq 1.4$ THz) and relative measurement complexity compared to the TPS. As such, the material characterization in a wide frequency range is produced using a set of BWO sources because the reliable spectral range of each BWO is limited to $\leq 100...200$ GHz. In addition, two sample and reference signals must be obtained to correctly reconstruct a sample complex refractive index using a BWO system. Conversely, TPS spectroscopy requires only one reference and sample waveform to obtain the complex transfer function of sample in a wide frequency range. Therefore, to determine the sample material parameters, TPS appears to be more convenient.

In order to compare TPS with the BWO spectroscopy techniques, several test samples were studied. The described TPS system as well as BWO spectrometer based on a Golay-cell detector were used for this purpose. Figure 1 shows the results of the test sample material parameter reconstruction: (a) shows the refractive indexes $n_1(\nu_t)$ and (b) shows the absorption coefficients $\alpha_1(\nu_t)$ of the test samples. The following three flats were studied: (1) is a highly absorbing 0.46 mm thick ceramic sample, (2) is a GaAs 1.19 mm thick flat, and (3) is a low-absorbing 1.37 mm thick sample. The described algorithm was implemented for the TPS waveforms processing, and the standard BWO spectroscopy data technique [21–23] was utilized. Continuous curves in Fig. 1 represent the TPS measurements, and the markers show the BWO spectroscopy results. Furthermore, the error bars indicate a $\pm 3\%$ uncertainty for the BWO spectroscopy measurements.

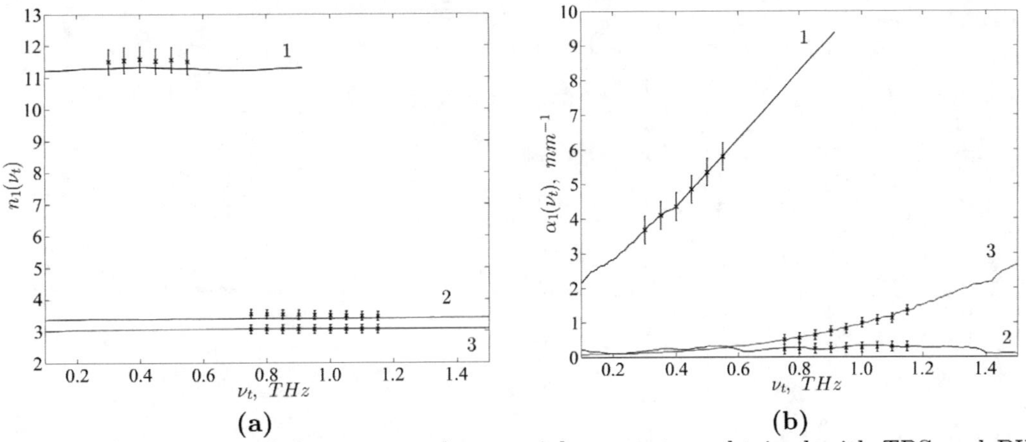

Figure 1. Comparison of the test-sample material parameters obtained with TPS and BWO spectroscopy: (a) shows the refractive indexes and (b) shows the absorption coefficients of (1) highly absorbing 0.46 mm-thick ceramic flat, (2) 1.19 mm-thick GaAs flat, and (3) low-absorbing 1.37 mm-thick ceramic flat.

One could note high accuracy of the material parameter reconstruction, since the TPS measurements results match the confidence interval of the BWO measurements. Accurate reconstruction was demonstrated for both high-absorbing (Fig. 1, curves 1) and low-absorbing (Fig. 1, curves 2) flats as well as for cases of satellite pulse appearance in the TPS waveforms (Fig. 1, curves 2 and 3) .

4. Conclusion

Novel results for the sample material parameter reconstruction using the TPS are demonstrated in the present work. Material parameter reconstruction procedure allowing studying the flat samples with small thickness or low refractive index, was realized and experimentally implemented for the test sample studying. Comparison of the reconstructed data with the result of the same sample studying by the use of the BWO spectroscopy helps both to verify the algorithm and to demonstrate the accuracy of the proposed technique.

Acknowledgments

This work was supported by the Russian Foundation for Basic Research (Project No.14-08-31102, No.14-08-31124, No.14-02-00781, and No.14-02-00256).

References

[1] Lee Y S 2009 *Principles of Terahertz Science and Technology* (New York, NY, USA: Springer)
[2] Pupeza I *et al.* 2007 *Opt. Express* **15** 4335
[3] Zaytsev K I *et al.* 2014 *J. Phys. Conf. Ser.* **486** 012018
[4] Zaytsev K I *et al.* 2014 *J. Appl. Phys.* **115**
[5] Zaytsev K I *et al.* 2013 *Opt. Eng.* **52** 068203
[6] Auston D H *et al.* 1978 *Solid State Electron.* **21** 147
[7] Mittleman D M *et al.* 1998 *App. Phys. B* **67** 379
[8] Jacobs E *et al.* 2012 *Opt. Eng.* **51** 091601
[9] Furxhi O *et al.* 2012 *Opt. Eng.* **51** 091612
[10] Perov A N *et al.* 2014 *J. Phys. Conf. Ser.* **486** 012027
[11] Woodward R M *et al.* 2003 *J. Invest. Dermatol.* **120** 72
[12] Wallace V P *et al.* 2004 *Br. J. Dermatol.* **151** 424
[13] Zaytsev K I *et al.* 2014 *J. Phys. Conf. Ser.* **486** 012014

[14] Taylor Z D *et al.* 2008 *Opt. Lett.* **33** 1258
[15] Arbab M H *et al.* 2011 *Biomed. Opt. Express* **2** 2339
[16] Ashworth P C *et al.* 2009 *Opt. Express* **17** 12444
[17] Reid C B *et al.* 2011 *Physics in Medicine and Biology* **56** 43334353
[18] Stoik C D *et al.* 2008 *Opt. Express* **16** 17039
[19] Yakovlev E V *et al.* 2014 *J. Phys. Conf. Ser.* **486** 012008
[20] Volkov A A *et al.* 1985 *Infrared Phys.* **25** 369
[21] Kozlov G V and Volkov A A 1998 *Millimeter and Submillimeter Wave Spectroscopy of Solids. Topics in Applied Physics* **74** 51
[22] Gorshunov B *et al.* 2005 *Int. J. Infrared Milli.* **26** 1217
[23] Gorshunov B P *et al.* 2013 *J. Phys. Chem. Lett.* **4** 2015
[24] Zhong H *et al.* 2006 *Opt. Express* **14** 9130
[25] Zhukova E S, *et al.* 2014 *J. Phys. Conf. Ser.* **486** 012019

Influence of raster scan parameters on the image quality for the THz phase imaging in collimated beam with a wide aperture

M S Kulya[1], N V Petrov[1], A N Tcypkin[1], V G Bespalov[1]
[1]Saint Petersburg National Research University of Information Technologies, Mechanics and Optics, Kronverkskiy pr. 49, St. Petersburg 197101, Russia

E-mail: Nickolai.petrov@gmail.com

Abstract. We numerically investigated the influence of wavelength of terahertz radiation and size of scanning pinhole on the quality of reconstructed image by terahertz pulse time domain holography method. The improvement of the quality is achieved mainly by reducing the wavelength of the radiation, pinhole size and pinhole step (value of the pinhole displacement at each step) affect the spatial resolution of the reconstructed object and significantly affect the scanning time. However, pinhole does not significantly affect the quality only if pinhole size is not less than used wavelength.

1. Introduction

The technologies of terahertz (THz) imaging are important for many applications [1] because they allow to obtain more information about the object of investigation [2-3]. Terahertz pulse time domain holography (THz PDTH) is one of the perspective methods allowing imaging of both amplitude and phase objects [4]. One of the features of THz PTDH is the scanning of an object in a wide collimated THz beam. Experimentally this method is based on direct measurements of the spatial and temporal distribution of the THz field and extracting the phase information using Fourier analysis procedure and further reconstruction the phase delay on the object. The purpose of this work is the numerical investigation of the influence of image reconstruction parameters on the quality of the reconstructed object. For the numerical evaluation of the quality of THz PTDH method we used modulation transfer function (MTF) [5]. We investigated the influence of the size of the scanning aperture and the wavelength of the THz radiation on the contrast of the reconstructed image.

2. Numerical investigation of image reconstruction quality by THz PDTH

We have performed a numerical investigation of impact of setup parameters, such as wavelength and size of scanning pinhole, on the quality of amplitude object reconstruction. We investigated the dependence of MTF of reconstructed object on the size of the scanning pinhole and on THz wavelength. The distance between the object plane and the registration plane was fixed at 70 cm, the object size was fixed at 20x20 cm, pinhole size changed from 400 μm to 200 μm, spectral range was from 400 μm to 200 μm. As a sample we used periodic grating with a line structure (Figure 1a). Figure 1 shows reconstruction of the sample at wavelength 400 μm, 300 μm and 200 μm in case of pinhole size 400 μm. For the reconstructed images we calculated MTF and plotted the 3D dependence of contrast characteristics of reconstructed object. These dependences are for different pinhole sizes when fixed wavelength (Figure 2) and for different THz wavelengths when fixed pinhole size (Figure 3). In doi:10.1088/1742-6596/536/1/012010

spectral range 200-400 μm pinhole size does not affect MTF significantly, but THz wavelength increases MTF.

3. Evaluation of a raster scanning time

THz wide collimated beam is raster scanned by square pinhole, which is fixed on the two coordinates controlled motorized translation stage. The motorized translation stage provides measurements of THz pulse temporal form at each step. Placing the object before the pinhole we obtained the spatial distribution of temporary forms of THz field, taking into account the phase shift during the passage of the THz field through the phase object.

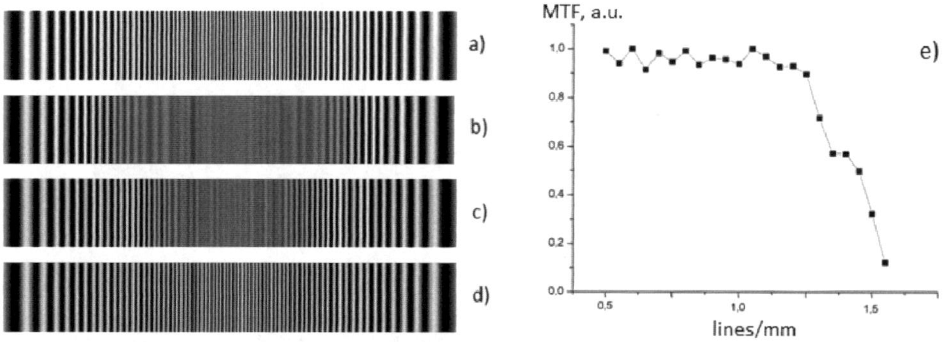

Figure 1. a)- cross section of initial object, b)- cross section of reconstructed object at wavelength 400 μm, c)- cross section of reconstructed object at wavelength 300 μm, d) cross section of reconstructed object at wavelength 200 μm, e) MTF function for (c).

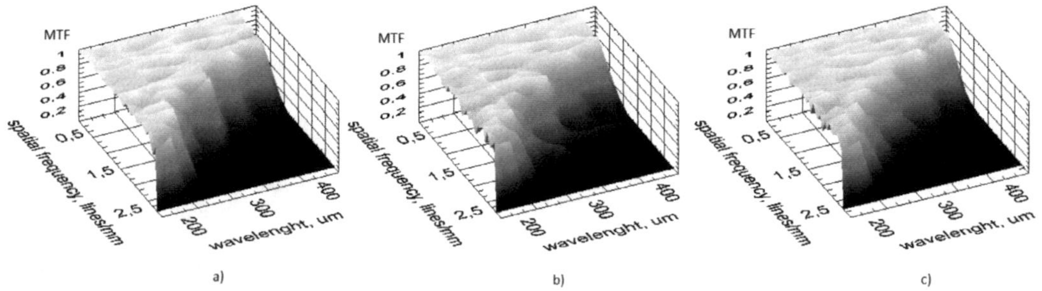

Figure 2. 3D dependence of MTF of reconstructed image from the wavelength when fixed pinhole size. a) pinhole size 200 μm, b) pinhole size 300 μm c) pinhole size 400 μm.

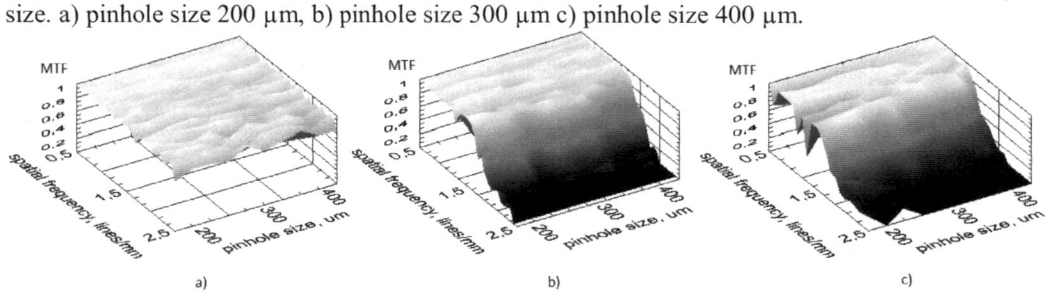

Figure 3. 3D dependence of MTF of reconstructed image from the pinhole size when fixed wavelength. a) 200 μm, b) 300 μm, c) 400 μm

In general, the time spent on scanning can be estimated as follows:

$$T_{scan} = \left(\left(\frac{L_z N_x N_y}{V_{z+}} + \frac{L_z N_x N_y}{V_{z-}}\right) + \left(\frac{\xi \cdot A_p \cdot N_x N_y}{V_{x,y}}\right) + T_{wait}\right) \cdot N_{mean} \tag{1}$$

where L_z - movement distance of translation stage by *z*, N_x - number of scans by *x*, N_y - number of scans by *y*, V_{z+} - forward movement speed of the translation stage by *z*, V_{z-} - backward movement speed of the translation stage by *z*, A_p - pinhole size, ξ - pinhole shift factor which determines the displacement of pinhole at each step relative to the pinhole size, $V_{x,y}$ - movement speed of the translation stage by *(x,y)*, T_{wait} - waiting time of the translation stage, N_{mean} - number of repetitions on each scanning step.

Reducing the raster scan time is possible when: 1) using involute stage [6] 2) reducing the number of steps in *(x,y)* scanning grid, 3) moving the translation stage in *(x,y)* grid while back movement by *z* axis, 4) reducing the temporal window for THz pulse by reducing the length of the displacement of the translation stage, however that leads to the loss of spectral information about the THz pulse, 5) increasing the speed of the translation stage, however it can reduce signal-to-noise ratio (SNR), 6) reducing the number of measurement repetitions at each scanning step, but it is only useful in case of a high SNR, 7) using single meaning measurement during slow movement by *z* instead of increasing the total number of repetitions N_{mean}, 8) using multichannel balanced electro-optic detection [7], or IR camera or array of microbolometers as a registration system.

4. Conclusion

We investigated parameters that affect the scanning time and the quality of image reconstruction. Based on these parameters we can optimize the experimental setup for THz PTDH. Improvement of MTF of reconstructed object has been achieved mainly by reducing the wavelength of radiation used in the image reconstruction. Pinhole size affects the spatial resolution of the reconstructed object and significantly affects the scanning time, but pinhole size does not affect MTF. However, in real experimental setup shorter wavelengths are limited by SNR. In practice, it is necessary to use more powerful sources with a wide spectrum. Also SNR is decreased by decreasing pinhole size. There is an optimal balance between scanning time and quality of image reconstruction. Scanning time is determined by the pinhole size, which is determined by necessary spatial resolution of the object and the SNR. The quality of image reconstruction depends on the wavelength of used THz radiation which is determined by the width of its spectrum.

Acknowledgments
This work was partially supported by RFBR Foundation (contract number 13 NK-02-013 458 \ 13 of 13.05.2012), financial support from Government of Russian Federation, Grant No 074-U01 and financial support from Russian Ministry of Education and Science project within the state mission for institutions of higher education (Agreement No 2014/190).

References

[1] Chan W L *et al* 2007 *Rep. Prog. Phys.* **70** 1325
[2] Chan W L 2008 *Opt. Lett.* **33** 974-976
[3] Zhang L 2009 *JOSA A* **26** 1187–1190
[4] Petrov N V *et al* 2013 *Proc. SPIE* **8846** 88460S
[5] Goodman J W 1968 *Infroduction to Fourier Optics* (New York: McGraw-Hill Book Co.)
[6] Xu J and Zhang X C 2004 *Opt. Lett.* **29** 2082-2084
[7] Pradarutti B *et al* 2007 *Opt. Express.* **15** 17652-17660

Measurement of characteristics and phase modulation accuracy increase of LC SLM "HoloEye PLUTO VIS"

A P Bondareva[1], P A Cheremkhin[1], N N Evtikhiev[1], V V Krasnov[1], R S Starikov[1], S N Starikov[1]

National Research Nuclear University MEPhI (Moscow Engineering Physics Institute), Faculty of Experimental and Theoretical Physics, Department of Laser Physics, Kashirskoe shosse 31, Moscow, Russia, 115409

E-mail: holo_mephi@mail.ru

Abstract. Phase liquid crystal spatial light modulators (LC SLM) are actively integrated in various optical systems for dynamic diffractive optical elements imaging. To achieve the best performance, high stability and linearity of phase modulation is required. This article presents results of measurement of characteristics and phase modulation accuracy increase of state of the art LC SLM with HD resolution "HoloEye PLUTO VIS".

1. Introduction

Phase liquid crystal spatial light modulators (LC SLM) [1] are actively integrated in various optical systems for dynamic diffractive optical elements imaging. However, majority of scientific applications [2-8] require stable phase modulation which might be hard to achieve with commercially available SLM due to its initial orientation on applications in consumer devices such as projectors.

State of the art LC SLMs implement digital voltage addressing scheme which simplifies controllers and allows to address high quantities of pixels ($\sim 10^7$). This, however, leads to phase temporal fluctuations due to step-wise voltage addressing. As shown in [9-10], voltage addressing frequency depends on quantity of addressable phase levels. Higher number of addressable levels requires lower frequency which in turn leads to higher amplitude of phase fluctuations.

In this paper phase LC SLM "HoloEye PLUTO VIS" was analyzed and tuned for the best performance. This SLM has the following characteristics: operation mode – reflection, quantity of pixels - 1920×1080, pixel size - 8×8 μm^2, frame rate - 60 Hz, quantity of addressable phase levels - 256 (8 bit).

2. Phase modulation measurement and accuracy increase

For SLM phase modulation measurements double-beam interference method was used [10-11]. Experimental setup is shown in Figure 1. The light source is HeNe laser with 633 nm wavelength. Laser beam is filtered and collimated with lenses L1, L2 and pinhole D1. Polarizer P selects horizontal polarization. Diaphragm D2 bounds beam aperture. Diffractive grating splits beam into three parts: zero and two first diffraction orders. Absorbing shield blocks zero diffraction order. Lens L3 forms image of diaphragm D2 in SLM plane. Analyzer A selects horizontal light polarization. Lens L4 in conjunction with lens L3 forms image of diffractive grating in camera photosensor plane. Due to blocked zero diffraction order image of diffractive grating has doubled spatial frequency. Actually this image is a result of interference of two first diffraction orders.

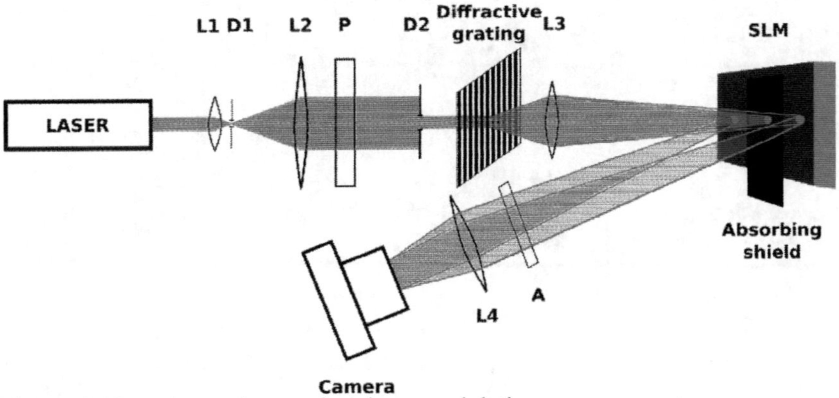

Figure 1. Experimental setup for phase modulation measurements.

Interference patterns were registered with scientific camera Megaplus II ES11000. While addressing the first half of SLM with different signal levels and the second one with zero signal, phase shift value changes accordingly, and thereafter location of interference fringes shifts. Location shift value corresponds to phase shift value. Three voltage addressing sequence configurations supplied with SLM were analyzed [10-11]: "18-6" (default configuration; allows for 1216 addressable phase levels), "5-5" (allows for 192 levels) and "0-6" (allows for 64 levels). Depending on quantity of addressable phase levels voltage addressing frequency changes accordingly. Less phase levels allows for higher voltage addressing frequency. For configuration "18-6" this frequency is equal to 120 Hz and for "0-6" it equals 480 Hz. Also, higher voltage addressing rate leads to lower maximum phase modulation depth.

Main characteristics of SLM for three measured configurations were derived. They are given in Table 1. Maximum momentary phase shift is maximum phase shift detected for all signal values and time delays. Maximum averaged phase shift is maximum phase shift averaged over frame duration for all signal values. Maximum phase fluctuations amplitude is half of maximum difference between the highest and the lowest phase shift for any given signal value. Quantity of distinguishable equidistant phase levels were determined as ratio of maximum phase shift value to doubled maximum phase fluctuations amplitude plus unity (because first and last phase levels have width equal to single amplitude).

Table 1. Characteristics of phase modulation of LC SLM "HoloEye PLUTO VIS".

Sequence configuration	"18-6"	"5-5"	"0-6"
Maximum momentary phase shift, π	3.17	3.01	2.39
Maximum averaged phase shift, π	3.05	2.88	2.29
Maximum phase STD, π	0.24	0.13	0.07
Maximum phase fluctuations amplitude, π	0.48	0.23	0.13
Quantity of distinguishable equidistant phase levels	4	7	10

With phase fluctuations increase, interference fringes visibility decreases. Measured dependencies of phase STD on signal value and interference fringes visibility on signal value for configuration "0-6", which provides lowest fluctuations, are shown in Figure 2.

It is clear that there is definite relation between values of phase STD and visibility. Higher fluctuations cause decrease of interference fringes visibility which is measure of coherence between modulated and unmodulated beams.

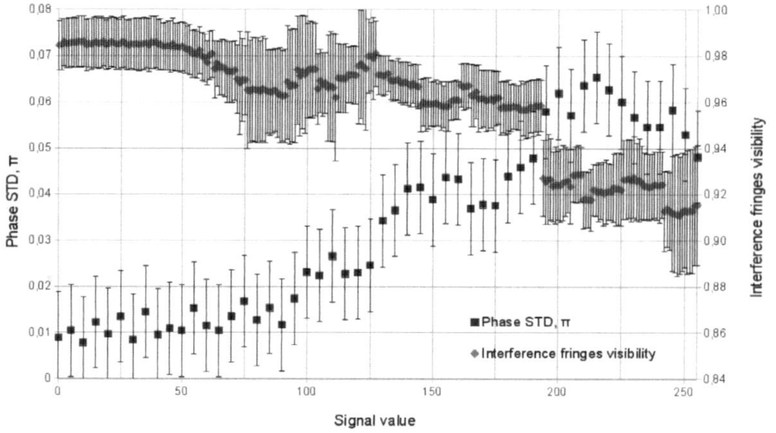

Figure 2. Dependencies of phase STD and interference fringes visibility on signal value for configuration "0-6".

3. Conclusion

Phase modulation characteristics of LC SLM "HoloEye PLUTO VIS" with three different voltage addressing sequence configurations were analyzed. Dependencies of phase shift on addressed signal value and time delay from frame start were obtained. It was determined that with default settings (configuration "18-6") SLM has phase STD equal to 0.24π and only 4 distinguishable equidistant phase levels. Highest characteristics were achieved with configuration "0-6": phase STD equals 0.07π and quantity of distinguishable equidistant phase levels equals 10. Definite relation between phase STD and interference fringes visibility was demonstrated.

Acknowledgments

This work was partially supported by the Russian Foundation for Basic Research (RFBR) № 13-07-00395.

References

[1] Wu S T and Yang D K 2005 *Reflective Liquid Crystal Displays* (Chichester: John Wiley)
[2] Schnars U and Juptner W 2002 Meas. Sci. Technol. **13** R85-R101
[3] Evtikhiev N N, Starikov S N, Cheryomkhin P A, Krasnov V V and Rodin V G 2012 *Proc. SPIE, Optical Modelling and Design II (Brussels)* vol 8429 (SPIE) p 84291M
[4] Kohler C, Schwab X and Osten W 2006 *Appl. Opt.* **45** 960-7
[5] Evtikhiev N N, Starikov S N, Cheryomkhin P A, Krasnov V V and Rodin V G 2012 *Proc. SPIE, Optical Modelling and Design II (Brussels)* vol 8429 (SPIE) p 84291P
[6] Wei J, Zhongyu C, Jacky W F, Changhe Z, Tak C Y and Sheun C P 2011 *Appl. Opt.* **50** H10-5
[7] Lee M, Gibson G, Bowman R, Bernet S, Phillips D and Padgett 2013 *Opt. Express* **21** 16541-51
[8] Cheremkhin P A, Evtikhiev N N, Krasnov V V, Porshneva L A, Rodin V G and Starikov S N 2014 *Proc. SPIE, Optical Modelling and Design III (Brussels)* vol 9131 (SPIE) p 913124
[9] Hermerschmidt S, Osten W, Krüger S and Blümel T 2007 *Proc. SPIE, Adaptive Optics for Laser Systems and Other Applications (Prague)* vol 6584 (SPIE) p 65840E
[10] Lizana A, Moreno I, Ruiz Márquez A, Iemmi C, Campos J and Yzuel M J 2009 *Proc. SPIE, Optics and Photonics for Information Processing III (San Diego)* vol 7442 (SPIE) p 74420Y
[11] Cheremkhin P A, Evtikhiev N N, Krasnov V V, Starikov S N and Rodin V G 2014 *Proc. SPIE, Practical Holography XXVIII: Materials and Applications (San Francisco)* vol 9006 (SPIE) p 900615

Application of speckle image correlation for real-time assessment of metabolic activity in herpes virus-infected cells

A P Vladimirov[1,2], A S Malygin[1], J A Mikhailova[1,2], E M Borodin[1], A A Bakharev[2] and A P Poryvayeva[2]
[1] Ural Federal University, Yekaterinburg, Mira street ,19, 620002, Russia
[2] FSSI Yekaterinburg Research Institute of Viral Infections of Rospotrebnadzor, Yekaterinburg, Letnaya Street, 23, 620030, Russia

E-mail: vap52@bk.ru

Abstract. Earlier we reported developing a speckle interferometry technique and a device designed to assess the metabolic activity of a cell monolayer cultivated on a glass substrate. This paper aimed at upgrading the technique and studying its potential for real-time assessment of herpes virus development process.
Speckle dynamics was recorded in the image plane of intact and virus-infected cell monolayer. HLE-3, L-41 and Vero cells were chosen as research targets. Herpes simplex virus-1-(HSV-1)-infected cell cultures were studied.
For 24 h we recorded the digital value of optical signal I in one pixel and parameter η characterizing change in the distribution of the optical signal on 10×10-pixel areas. The coefficient of multiple determination calculated by η time dependences for three intact cell cultures equals 0.94. It was demonstrated that the activity parameters are significantly different for intact and virus-infected cells. The difference of η value for intact and HSV-1-infected cells is detectable 10 minutes from the experiment start.

1. Introduction

Human diseases are directly or indirectly associated with cell function impairment. That is why a comprehensive view of normal and pathological cell function mechanisms is vital for both rapid diagnosis of diseases and timely administration of medicines in every individual. Recording laser speckle, or biospeckle, dynamics [1] is a promising tool for studying microscopic processes in biological media. Speckle (a term of an English origin) is a random interference pattern created by mutual interference of multiple coherent waves with random phase shifts. When live objects are illuminated by laser the speckle pattern changes due to temporal changes of scattered wave phases and amplitudes. Speckle dynamics was used to assess the biological activity of seeds [2], fruits [3] and other objects. In [4] speckle contrast imaging was used to assess patients' blood flow in retinae and skin-supplying capillaries of their extremities.

Earlier, a speckle interferometry technique and a device for assessing the metabolic activity of cultivated cells were developed using L-41 cells [5]. The optical difference variation value of the object-probing wave pair path σ_u was selected as a cell activity parametre. High labour input precluding real-time measurements was the disadvantage of the technique.

The objective of this research was upgrading the technique and studying its potential for real-time assessment of herpes virus development process.

2. Experimental procedure

Speckle dynamics was recorded in the image plane of the intact and virus-infected cell monolayer. L-41 CD/84, HLE-3 and Vero cells were selected as the research targets. Herpes simplex virus-1(HSV-1)-infected cell cultures were studied. Experiments were conducted using the device represented in Figure 1.

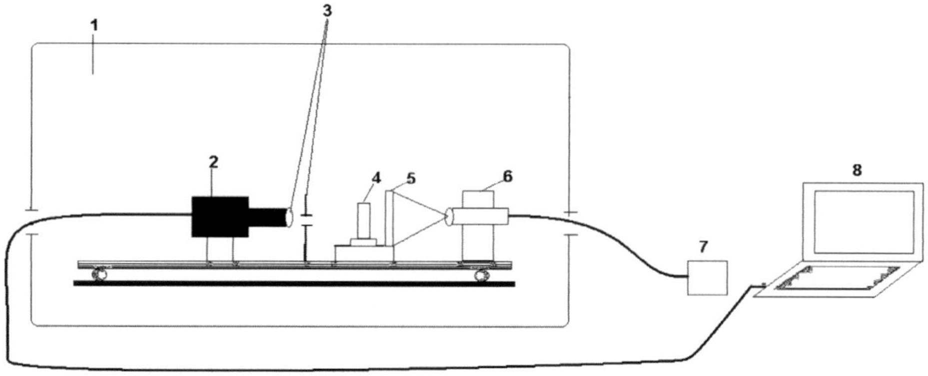

Figure 1. Optical device scheme: 1 – thermostat, 2 – telecamera, 3 – lens with a diaphragm, 4 – tray, 5 – opal glass, 6 – laser module, 7 – line adapter (5V), 8 – PC

For 24 h, we recorded the digital value of optical signal I in one pixel and parameter η characterizing change in the distribution of the optical signal on 10x10-pixel areas. The TV camera exposure time was 9 s, exceeding the radiation intensity correlation time (5-8 s) calculated earlier in [5].

η value was determined using formula (1):

$$\eta = \frac{\frac{1}{m\cdot n}\sum_{i=0}^{m-1}\sum_{j=0}^{n-1}[A_{i,j}-\bar{A}][B_{i,j}-\bar{B}]}{\left(\frac{1}{m\cdot n}\sum_{i=0}^{m-1}\sum_{j=0}^{n-1}[A_{i,j}-\bar{A}]^2\right)^{1/2}\cdot\left(\frac{1}{m\cdot n}\sum_{i=0}^{m-1}\sum_{j=0}^{n-1}[B_{i,j}-\bar{B}]^2\right)^{1/2}}, \tag{1}$$

where $A_{i,j}$ is the digital value of optical signal I in $m\times n$-pixel area at the reference time, $B_{i,j}$ is the digital value of optical signal I in the same area after time τ, I and j are ordinal pixel numbers in the directions of the x and y axes, respectively, $\bar{A}$ is the mean digital value of optical signal I at the reference time, and $\bar{B}$ is the mean digital value of optical signal I after time τ.

Figure 2 represents time dependences of parametre η for HSV-1-infected and intact Vero cells.

Value η tends to decrease gradually for both intact and virus-infected cells. For HLE-3 and L-41 cells, similar dependences were observed. The coefficient of determination (for comparing HSV-1-free and HSV-1-infected cells) was 0.934 for HLE, 0.89 for Vero, and 0.87 for L-41. For three intact cell cultures the coefficient of multiple determination was 0.939. The coefficient value around 1 suggests similarity of the process types in the cells changing the paths of scattered waves. The experimental data analysis showed that the difference of value η for intact and HSV-1-infected cells is detectable 10 minutes from the experiment start.

3. Conclusion

The studies performed may lead to the following conclusions:

1) For various HSV-1-infected and intact cell cultures the time dependences of η have the same trend: value η decreases nonlinearly with increasing experimental time. The coefficient of multiple determination calculated using the dependences $\eta = \eta(\tau)$ for three intact cell types equals 0.94. The coefficient of determination for intact and virus-infected cell cultures is 0.934 for HLE, 0.89 for Vero and 0.87 for L-41 cells.

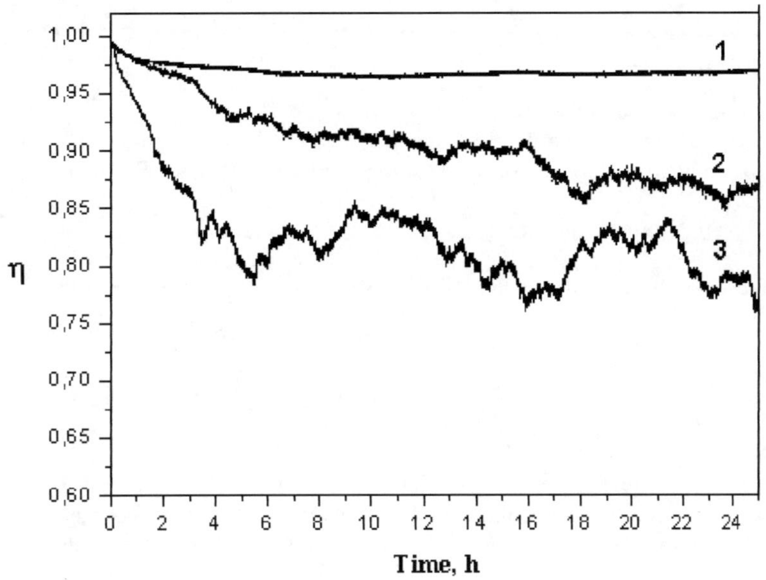

Figure 2. Time dependence of η for Vero cells: 1 – growth medium, 2 – intact cells, 3 – HSV-1-infected cells

2) Time dependences of η for virus-infected and intact cell cultures differ significantly. The difference in value η for intact and HSV-1 infected cells is detectable 10 minutes from the experiment start.

References

[1] *Dynamic Laser Speckle and Applications* 2008 ed Hector J Rabal and Roberto A Braga Jr (New York: CRC Press)

[2] Cardoso R R, Costa A G, Nobre C M B and Braga R A 2011 Frequency signature of water activity by biospeckle laser *Optics Communications* **284** 2131–36

[3] Oulamara, Tribillon G and Duvernoy J 1989 Biological activity measurement on botanical specimen surfaces using a temporal decorrelation effect of laser speckle *Journal of Modern Optics* **36** 165–79

[4] Briers J D 2007 Laser speckle contrast imaging for measuring blood flow *Optica Applicata* **XXXVII** 139-52

[5] Malygin A S, Bebenina N V, Vladimirov A P, Mikitas K N and Baharev A A 2012 A speckle-interferometric device for studying the cell biological activity *Instruments and Experimental Techniques* **55** 415–18

Curvature histogram features for retrieval of images of smooth 3D objects

I Zhdanov[1], O Scherbakov[1], A Potapov[1], M Peterson[1]
[1] Saint Petersburg National Research University of Information Technologies, Mechanics and Optics, Kronverkskiy pr. 49, Saint Petersburg 197101, Russia

E-mail: avenger15@yandex.ru

Abstract. We consider image features on the base of histograms of oriented gradients (HOG) with addition of contour curvature histogram (HOG-CH), and also compare it with results of known scale-invariant feature transform (SIFT) approach in application to retrieval of images of smooth 3D objects.

1. Introduction

The difficulty of image retrieval by text-based indexing significantly increases when one deals with images of smooth 3D objects. A typical example of such object is a modern abstract sculpture. A potential user, who would like to know the name and the author of sculpture, most probably will be confused to describe the shape of the object observed in the image. Thus, the search by text description is hardly applicable here and image retrieval should be done in a fully automatic mode without prior information about observed objects. In this case we deal with content-based image retrieval alone, i.e. we need to build an image descriptor automatically without any user interaction and provide a matching algorithm in order to find a corresponding object in the image base.

Direction of intensity gradient vector is a strong discriminative feature for images of smooth 3D objects, and one can use it for image retrieval. Our research was mainly inspired by work of Arandjelovic and Zisserman [1], where they presented a bag of boundaries (BoB) approach, based on histograms of oriented gradients (HOG) [2], the gPb contour detector [3] and prior segmentation. Two last steps are highly time-consuming, so here we tested less sophisticated technique with a simpler contour detection algorithm on the base of classical Deriche filter [4] and without prior segmentation, which can provide incorrect results for complicated scenes. The main motivation was in further implementation of retrieval system on a mobile platform. Besides this, we investigate a variation of HOG features which combines histograms of oriented gradients and curvature histogram (HOG-CH). Finally, descriptors on the base of scale-invariant feature transform (SIFT) approach [5] was also examined for comparison.

2. Image features evaluation

Descriptor of image consists of the set of feature vectors with N elements which contain information about low-level features, for instance, shape, color, texture, *etc*. It is a convenient way of image representation, storage and matching. Thus, image retrieval task can be formalized as searching for the closest feature vector with N elements with respect to a specified metric, for example, Euclidean distance. The more matched features we have, the more images are similar to each other.

Journal of Physics: Conference Series **536** (2014) 012013 doi:10.1088/1742-6596/536/1/012013

Here we evaluate features on the base of histogram of gradient orientations. The first step is Deriche's filtration of grayscale image with further edge detection and tracking. As a result we have a bunch of edges with known coordinates of each pixel belonging to the edge along with gradient magnitude and orientation at that coordinates. After selecting edges with appropriate length we sample key points at regular interval and compute feature vector for each of them. Following [1], feature vectors are computed at multiple scales in order to represent boundary information locally. The area of region of interest around selected point on the edge is set to be 1/30, 2/15 and 8/15 of the image area. Every patch is scaled to 32×32 elements, thus for computation of HOG we use 4×4 cell array, with each cell containing 8×8 elements. The HOG feature vector has 324 elements which are computed via spatial binning into 9 gradient orientations for 9 overlapping blocks each with 2×2 cells within scaled patch. The difference of our approach from HOG is that we compute histograms only for points on the edge, thus features becomes more stable to a possible background variation.

The second part of the HOG-CH feature vector includes information about curvature of the edge at selected point. Curvature parameter can be computed using formula (1):

$$C = |L|/(S_l+S_r), \quad (1)$$

where $|L|$ – length of the line L between neighboring pixels from left and right side of edge with respect to the current pixel, S_l and S_r are areas of geometric shapes that are formed by edge and line L (see Figure 1).

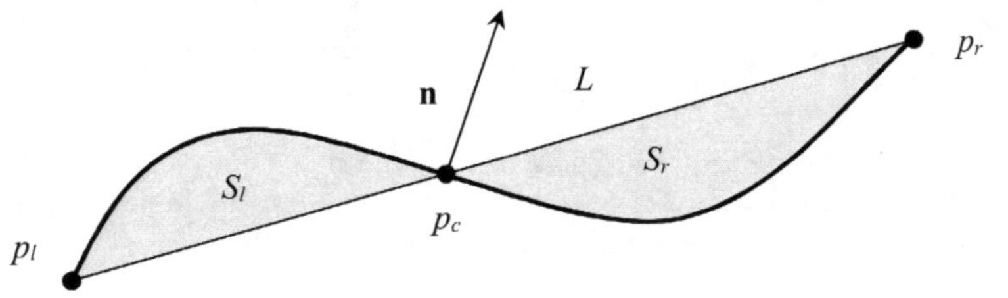

Figure 1. Explanation of curvature parameter. The line L drawn through p_c, p_l and p_r forms two geometrical shapes with areas S_l and S_r, **n** is an orientation vector, normal to the contour at the current pixel p_c, and p_l, p_r are correspondent left and right neighbors of the current pixel.

We can consider a line L as a rectangle with side equal to 1 and $|L|$. For a curved edge the sum of areas S_l and S_r will be bigger than the area of that rectangle, i.e. length of line L, so parameter C will be less than 1. If C is equal to 1, we will deal with a straight contour line. Final HOG-CH feature vector is formed as a concatenation of HOG part and histogram of C values.

3. Experimental results

For experimental evaluation we used Oxford's "Sculptures 6k" dataset with more than 6000 images available at [6]. HOG-CH, HOG and SIFT features were computed for every image in the dataset and formed a bank of image features. In the retrieval process we extracted features in the query image and matched them with features in the bank using the nearest neighbor scheme with additional constraints on the base of epipolar geometry. If the number of matched features was higher than defined threshold, we considered images as similar. Thus, image retrieval procedure consisted in searching for the correspondent image with maximal number of correctly matched features.

Figure 2 shows an example of retrieval results for HOG and HOG-CH features, and Figure 3 represents obtained statistics for 10 best matches. In total, HOG features gave approximately 78% of correct retrievals within first 10 results with the highest number of matched features, HOG-CH features provided 72%, while SIFT showed only 50% of correct retrievals.

Figure 2. Example of image retrieval results: (a) query images, (b) three best matches for HOG features, (c) three best matches for HOG-CH features.

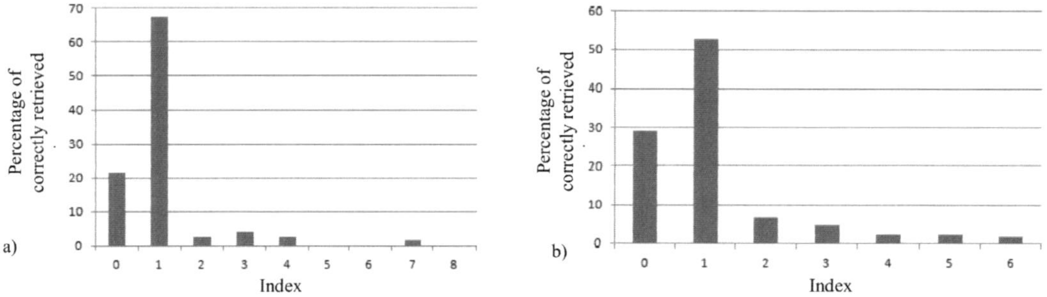

Figure 3. Statistics for retrieval resuts: (a) HOG features; (b) HOG-CH features. Horizontal axis represents index of the first correct retrival within 10 images with the highest number of matched features. Absense of correctly matched images is marked with index 0.

4. Conclusion

We have considered HOG-CH features in application to retrieval of smooth 3D object images. The features represent as a simplified modification of HOG with additional curvature histogram. Evaluated experiments showed that proposed features provide lower percentage of correctly retrieved images than HOG features, but exceed SIFT features significantly. HOG features give good results for images with textured background, because of consideration of gradient orientations within areas around pixels on the contour, while HOG-CH features are concentrated on the contour itself and show better result for images with homogeneous background. Thus, the presented approach is suitable for content-based retrieval of images of smooth 3D objects, but requires further enhancement in order to increase quality of image retrieval system.

Acknowledgements

This work was supported by the Russian Federation President's grant Council (MD-1072.2013.9) and the Ministry of Education and Science of the Russian Federation.

References

[1] Arandjelovic R, Zisserman A 2011 *Proc.of the IEEE ICCV* 375–382
[2] Dalal N, Triggs B 2005 *Proc.of the IEEE Conf. on CVPR* **1** 886–893
[3] Maire M, Arbeláez P, Fowlkes C, Malik J 2008 *Proc.of the IEEE Conf. on CVPR* 1–8
[4] Deriche R 1987 *International Journal of Computer Vision* **1** 167–187
[5] Yang J, Jiang Y G, Hauptmann A G, Ngo C W 2007 *Proc. of the International Workshop on Multimedia Information Retrieval* 197–206
[6] *Oxford's VGG "Sculptures 6k"* dataset http://www.robots.ox.ac.uk/~vgg/data/sculptures6k/

The study of plant tissue by digital microscopy methods

V V Chirskaya[1], N B Margaryants[1], E V Zhukova[1]
[1] Saint Petersburg National Research University of Information Technologies, Mechanics and Optics, 49 Kronverkskiy pr., Saint Petersburg 197101, Russia

E-mail: fosp@grv.ifmo.ru

Abstract. Results of the study of potato tuber (*Solanum tuberosum L.*) periderm with the use of digital microscopy methods are considered. A light microscope was used for diagnostics of periderm surface and recording of the images of studied sections. High-resolving full-field optical coherence microscopy ensured obtaining tomograms images of periderm texture in depth. The scanning depth was 32 μm. Tomograms (B-scans) of surfaces with normal structure of periderm upper layer in the area of mechanic damage of external protective tissues and also in the area of growth caused by a fungal disease are presented. Tomograms resolution by depth and in lateral directions is 1 μm.

1. Introduction

Optical coherence microscopy is a modern method of investigation the microstructure of various materials and substances. When studying living biological tissues, the employment of this method ensures the nondestructive nature of experimental work and creates conditions for visualization of tissue texture with the resolution that is required for diagnostics. Traditionally, this method is used to study the texture of biological tissues in order to advance providing diagnostic medical assistance and scientific study of biological processes of development and degradation of cellular and tissue structures. Plant tissues are also widely studied using optical coherence microscopy. For instance, the results of tradescantia leaves cells visualization are discussed in [1] and research on the influence of viral infections on the change of orchid tissue is described in [2]. The progress of studies conducted with the help of optical coherence microscopy methods depends on device resolution, software speed and effectiveness of computation algorithms when working with large data volumes, and, also, on the method of presenting information obtained for tissue texture on micro level. This article deals with the results of study of the potato tuber (*Solanum tuberosum L.*) periderm. The purpose of the work was to make a comparative study of periderm tissue for various types of its surface damages.

2. High-resolving full-field optical coherence microscope

The experiment was performed using a high-resolving full-field optical coherence microscope, whose structure combines the advantages of a low-coherence Michelson interferometer with Linnik configuration and the light microscope [3]. A halogen lamp with capacity of 70 W served as a broadband light source. A beam-splitting plate formed two parallel coherent beams, which were focused respectively on the object surface and reference mirror with the help of two micro objectives that have similar characteristics. Radiation reflected from the object and the mirror was collected by micro objectives and two coherent beams again went through the beam divider. Further, with the help of the lens, the beams were brought together on a plane of video-camera matrix, which is used for the registration of the observing interference pattern. The optical compensator was placed in the channel of reference beam, which allowed to bring in additional path difference for improving the contrast of interference patterns. The microscope magnification was 500^X and the area of the observation surface was 200x200 μm^2.

In experiment the stepper motor through fixed interval (6.4 nm) shifted the micro objective by depth with regard to the surface of the object under study. The optical path of the reference beam always remained the same. Owing to small shifts of the micro objective and small sizes of optical inhomogeneities, the conditions for observing the interference effect were met. At the specified scanning depth, an optical path difference between subject and reference beams for a number of waves was less than the coherence length of this type of source. A total interference pattern was observed at the focal plane of a lens. When changing the scanning depth, the conditions of interference observation were carried out for another waves.

Recording of video frames with the images of interference patterns was carried out jointly with the changing of the scanning depth through the equal intervals (64 nm). During the experiment, a depth of scanning was limited by 32 µm. USB 2.0 digital video-camera Color VEC-135 (1.3 MP) was used to record interference patterns, and registration of the interferometric images was carried out in the visible area of the spectrum. The size of the recording image was 1280x500 pixels. Combined computer processing of the array of the interference images (500 video-frames) with the help of a specially developed algorithm allowed to obtain the images of plant tissue tomograms (B-scans) for selected sections within the area of object surface observed with the microscope.

As a result of research, the tomograms of potato tuber periderm were obtained. The size of B-scans was 32 µm in height (scanning depth) and 200 µm in width (scanning length). The resolution of tomogram image by depth and in lateral directions was not more than 1 µm. Such scanning parameters allowed to examine the upper layers of potato tuber periderm.

3. Results

The tomogram of the periderm surface without defects is presented in Figure 1. The image provides normal layered tissue texture. For comparison, Figure 2 presents the tomogram obtained in the growth area, formed on periderm surface during potato tuber growth under the influence of plant pathogenic fungi. Such growths are formed in case of tissue infection with the fungi of *common potato scab*. The tomogram shows the profile of growth in the selected section. It is seen that the periderm tissue loses layered cellular texture.

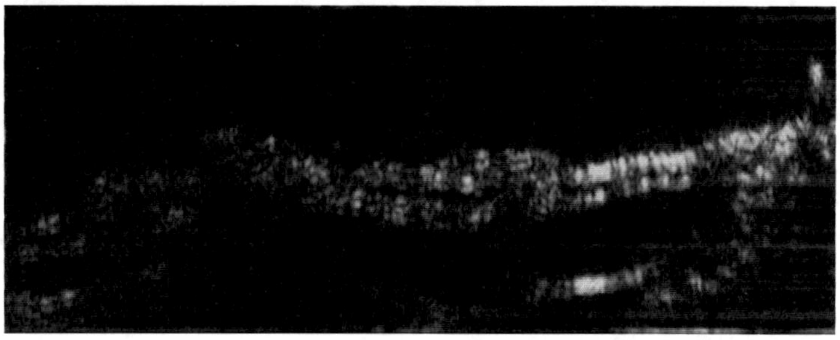

Figure 1. Tomogram of potato periderm without defects

The height of the growth at the section under study was about 17 µm, the length of its base was within the limits of 149 µm. Such estimation can be helpful for understanding the nature and possible tendencies in the defect development allowing the specialists to diagnose the character and the speed of tuber disease.

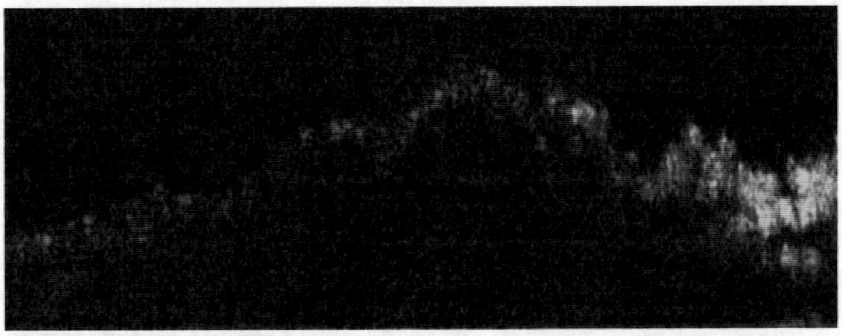

Figure 2. Tomogram of potato periderm surface in the growth area

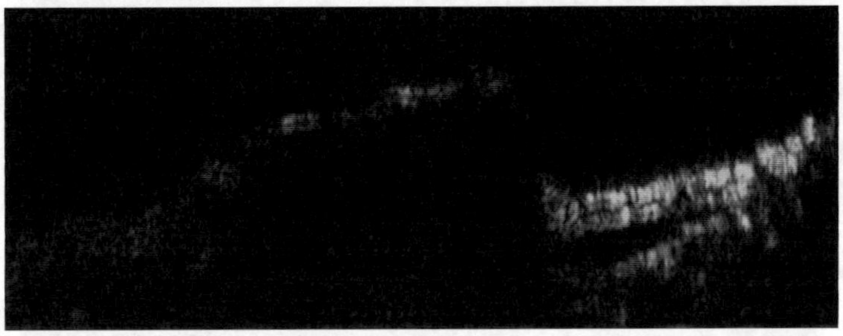

Figure 3. Figure 2. Tomogram of periderm with mechanic damage

The tomogram of periderm section surface with mechanic damage is presented in figure 3.It is seen that the rupture in the external protective tissue also causes a local breaking of the layered tissue texture. Also, it can be seen that the area of displacement of external tissue layer corresponds to the local breaking of its structural elements. The surface relief form points to the absence of dangerous changes in the preiderm texture.

4. Conclusions

The combination of two high-resolution digital optical methods for diagnostics of diseases and assessment sizes of periderm surface defect may be useful for investigations of other plants' periderm. The study of periderm by the high-resolving full-field optical coherence microscopy method showed that in order to observe the microstructure of layers located deeper from surface, it is necessary to use a more powerful source with emission in the near IR spectrum area and greater scanning depth. Also, in order to enhance the conditions of plant texture visualization, it is necessary to carry out additional studies of spectral characteristics of individual tissue textures of the object. We believe that the results of the study can be helpful to a wide range of specialists.

Acknowledgments
This work was financially supported by Government of Russian Federation, Grant 074-U01.

References

[1] Sapozhnikova V V, Kamenskii V A and Kuranov R V 2003 *Rus. Plant Physiol.* **50** 282
[2] Chow T H, Tan K M, Hg B K, Razul S G, Tay C F, Chia T F and Poh W T 2009 *J. Biomed. Opt* **14** 014006
[3] Alarousu E, Gurov I, Kalinina N, Karpets A, Margariants N, Myllylä R, Prykäri T and Vorobeva E 2008 *Proc. SPIE of Advanced Laser Technologies* (Finland, Levi, 03 September 2007) 702212

The second order extended Kalman filter and Markov nonlinear filter for data processing in interferometric systems

P Ermolaev[1] **and M Volynsky**[1]
[1] Saint Petersburg National Research University of Information Technologies, Mechanics and Optics, 49 Kronverkskiy pr., Saint Petersburg 197101, Russia

E-mail: petr-ermolaev@hotmail.com

Abstract. Recurrent stochastic data processing algorithms using representation of interferometric signal as output of a dynamic system, which state is described by vector of parameters, in some cases are more effective, compared with conventional algorithms. Interferometric signals depend on phase nonlinearly. Consequently it is expedient to apply algorithms of nonlinear stochastic filtering, such as Kalman type filters. An application of the second order extended Kalman filter and Markov nonlinear filter that allows to minimize estimation error is described. Experimental results of signals processing are illustrated. Comparison of the algorithms is presented and discussed.

1. Introduction

Interferometric methods are the most accurate among other optical methods [1, 2]. Signals obtained in interferometric systems are affected by noise. To recover information about properties of objects it is necessary to use the data processing algorithms robust to different types of noise [2].

Conventional processing methods based on Fourier transform do not consider available *a priori* information about a model of signal formation and need in acquisition of full signal realization before calculating that limits the processing speed. Recurrent stochastic data processing algorithms based on the state-space approach and formalism of stochastic differential equations perform consequent calculating of signal samples series. In this way, interferometric signals are represented as output of a dynamic system [2]. The most popular state-space estimation algorithm is linear Kalman filter (LKF). This filter is optimal in terms of minimum estimation error variance [3]. Dependence of measured interferometric signal on parameters of model is nonlinear, it is why the LKF is not applicable for such systems. To solve this problem it is necessary to use the algorithms of nonlinear stochastic filtering. Extended Kalman filter (EKF) [2, 4], the most widely used nonlinear Kalman type algorithm, uses linearization of nonlinear system equations by the first-order Taylor series expansion. To minimize approximation error one can use the second-order Taylor series expansion of state and measurement equations, which is used in the second order Kalman filter (SOEKF) [4] and Markov nonlinear filter (MNLF) [5]. In the paper, the EKF, SOEKF and MNLF algorithms that were applied to interferometric signals processing are compared and discussed.

2. Theoretical background

One can represent the state of a dynamic system by vector of parameters. In case of the interferometric signal, the vector of parameters includes background, amplitude and phase. Output of the system is measured discrete signal value.

The discrete dynamic system can be described by system and observation equations [3, 4]:

$$\boldsymbol{\theta}(k) = \mathbf{f}(\boldsymbol{\theta}(k-1)) + \mathbf{w}(k), \quad (1)$$

$$\mathbf{s}(k) = \mathbf{h}(\boldsymbol{\theta}(k)) + \mathbf{n}(k), \quad (2)$$

where $\boldsymbol{\theta}(k)$ is vector of parameters; $\mathbf{s}(k)$ is output of the dynamic system; $\mathbf{f}(\cdot)$ and $\mathbf{h}(\cdot)$ are nonlinear differentiable functions; $\mathbf{w}(k)$ is system noise; $\mathbf{n}(k)$ is observation noise; $k = 0..K{-}1$. System equation determines vector of parameters at each discrete-time sample. Observation equation (2) produces output signal by current vector of parameters. This equation is nonlinear because of the output depends on the signal phase nonlinearly.

Data processing by EKF, SOEKF, and MNLF is divided into two steps: prediction and update. In the first step, current vector of parameters is predicted using estimation of this vector at previous discrete-time sample taking into account the system equation. At the second step the gain coefficients of the filter are computed according to approximation of nonlinear equations and stochastic properties of noise. Predicted vector of parameters is updated by using measured value of the signal on current sample and the gain of the filter. In a manner, the gain determines degree of confidence to current measurements. Updated vector is the estimate of the vector of parameters at current sample.

Approximation of nonlinear system equations by the second-order Taylor series expansion in SOEKF and MNLF is carried using additional modifications of the prediction and update formulas, which are different for proposed algorithms. The prediction and update equations are changed in SOEKF compared with EKF by addition of terms using the second derivatives of nonlinear system equations. The gain in SOEKF is calculated similarly to the EKF. MNLF uses prediction of vector of parameters like in the EKF, but the equations for gain computation and update of predicted vector differ have differences [2, 4]. Table 1 illustrates the equations, which are different for described algorithms.

Table 1. Comparison of the algorithms

	EKF	SOEKF	MNLF
Prediction	$\hat{\boldsymbol{\theta}}(k) = \mathbf{f}(\boldsymbol{\theta}(k-1))$	$\hat{\boldsymbol{\theta}}(k) = \mathbf{f}(\boldsymbol{\theta}(k-1)) + \underline{\mathbf{A}(k)}$	$\hat{\boldsymbol{\theta}}(k) = \mathbf{f}(\boldsymbol{\theta}(k-1))$
Update	$\boldsymbol{\theta}(k) = \hat{\boldsymbol{\theta}}(k) +$ $+ \mathbf{P}(k)[\mathbf{s}(k) - \mathbf{h}(\hat{\boldsymbol{\theta}}(k))]$	$\boldsymbol{\theta}(k) = \hat{\boldsymbol{\theta}}(k) +$ $+ \mathbf{P}(k)[\mathbf{s}(k) - \mathbf{h}(\hat{\boldsymbol{\theta}}) - \underline{\mathbf{B}(k)}]$	$\boldsymbol{\theta}(k) = \hat{\boldsymbol{\theta}}(k) +$ $+ \mathbf{P}(k)[\mathbf{s}(k) - \mathbf{h}(\hat{\boldsymbol{\theta}}) - \underline{\mathbf{B}(k)}]$
Gain	$\mathbf{P}(k) = \mathbf{R}(k)\mathbf{H}(k)^{\mathrm{T}} \times$ $\times \left(\mathbf{H}(k)\mathbf{R}(k)\mathbf{H}(k)^{\mathrm{T}} + \mathbf{R}_n\right)^{-1}$	$\mathbf{P}(k) = \mathbf{R}(k)\mathbf{H}(k)^{\mathrm{T}} \times$ $\times \left(\mathbf{H}(k)\mathbf{R}(k)\mathbf{H}(k)^{\mathrm{T}} + \mathbf{R}_n\right)^{-1}$	$\mathbf{P}(k) = \mathbf{R}(k)\mathbf{H}(k)^{\mathrm{T}} \times$ $\times \left(\mathbf{H}(k)\mathbf{R}(k)\mathbf{H}(k)^{\mathrm{T}} + \mathbf{R}_n + \underline{\mathbf{C}(k)}\right)^{-1}$

In this table, $\mathbf{R}(k)$ is prediction of the covariance matrix of errors; $\mathbf{R}_n$ is covariance matrix of system noise; $\mathbf{H}(k)$ is observation matrix calculated by linearization of the observation equation. Underlined terms $\mathbf{A}(k)$, $\mathbf{B}(k)$, and $\mathbf{C}(k)$, which additionally use the second partial derivatives of equations (1) and (2) [2, 4, 5], represent difference between the algorithms. Computation of these terms requires additional operation time with respect to the computational time of the EKF algorithm.

3. Results

Figure 1 shows experimental results of interferometric signal processing by the MNLF algorithm (as an example). Signal amplitude varies in accordance with normal distribution with expected value at the 500th sample. Phase of signal is determined as $\Phi(k) = 2\pi(ak - bk^2)$, where $a = 10^{-1}$, $b = 200^{-2}$. The signal is affected by additive Gaussian noise with standard deviation equal to 5% of maximal amplitude. Phase variations are expected *a priori* as linear, i.e. $\Phi(k) = 2\pi ak$, where $a = 70^{-1}$. It has been found that the distribution of phase estimation error is close to Gaussian function. Maximal phase estimation error did not exceed $\pi/6$ rad.

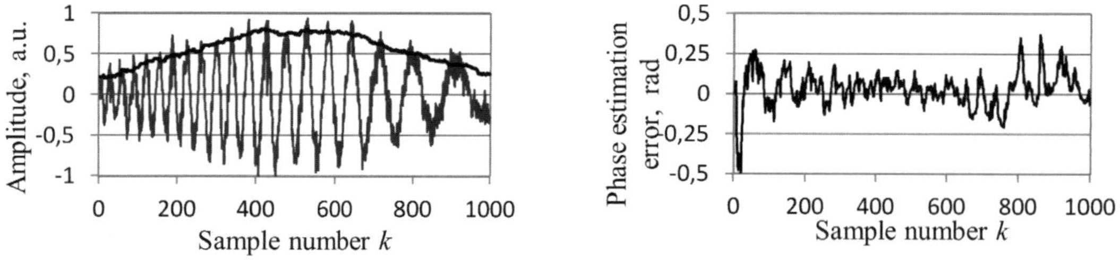

Figure 1. The results of signal processing by MNLF: the normalized signal and estimated amplitude (left plot) and phase estimation error (right plot).

Table 2 shows the obtained RMS errors of signal parameters and signal-to-noise ratio (SNR) of the restored signal. Initial SNR is equal to 10 dB.

Table 2. Standard deviations of parameters estimation errors of the algorithms

	EKF	SOEKF	MNLF
Background, %	1.11	1.12	1.10
Amplitude, %	4.45	4.46	4.23
Phase, rad	0.14	0.14	0.15
SNR, dB	17.90	17.92	17.84

The SOEKF and MNLF estimation errors differ little from the EKF estimation error. It shows that additional terms of the SOEKF and MNLF have small values in this case. Consequently using the high-order Taylor series expansion of nonlinear equations is not expedient for interferometric systems.

4. Conclusion

Considered algorithms are applicable for parameters estimation of interferometric signals with original SNR less than or equal to 10 dB. Typically maximal phase estimation error does not exceed $\pi/6$ rad. Comparison of the algorithms shows that quality of estimation by the algorithms, using the second-order Taylor series expansion of nonlinear system equations, differs little from the EKF using linearization for the considered kind of signal. Therefore, using of EKF for interferometric data processing is enough in terms of computational power minimization.

Acknowledgment

This work was fulfilled under financial support from the Government of Russian Federation (Grant 074-U01) and grant of the President of the Russian Federation for state supporting of young Russian PhD scientists (Grant MK-1455.2014.8).

References

[1] Hariharan P 2007 *Basics of interferometry* (San Diego: Elsevier)
[2] Gurov I and Volynsky M 2012 *Optics and Lasers in Engineering* **50** 514–521
[3] Kalman R 1960 *Trans. ASME, J. Basic Eng.* **82** 35–45
[4] Simon D 2006 *Optimal state estimation* (New-York: John Wiley & Sons, Inc.)
[5] Gurov I Sheynihovich D 2000 *JOSA A.* **17** pp 21–27

Concept-based lexical-semantic unsupervised learning system

V I Filatov[1]
[1]Saint Petersburg National Research University of Information Technologies, Mechanics and Optics, 49 Kronverkskiy pr., Saint Petersburg 197101, Russia

E-mail: sigal89@mail.ru

Abstract. Concept learning is essential for automatic knowledge acquisition and consists in linking semantic and linguistic data together. In this work, a system of concept learning, which mimics to some extent human infant learning, is presented. This system performs visual and audial feature extraction and construction of concept database by maximizing of their mutual information. Experiments show 90% recognition rate of learnt concepts.

1. Introduction

Within the history of artificial intelligence system engineering, scientists have proposed various ways of machine learning. At first, the best-known kinds of systems operating with knowledge were logic-based. Expert systems were based on binding existing knowledge in order to assist humans in decision-making. The next big breakthrough in intelligent systems consisted in development of automatic knowledge acquisition methods.

Most promising type of machine learning in knowledge acquisition is unsupervised learning since it doesn't require manually labeled datasets. Unsupervised learning systems are now very popular, and novel approaches are appearing every day. Some of them imitate animal reflexes [1].

This work describes novel algorithms and approaches to concept-based learning. An experimental computer system called Automatic Lexical-Semantic Artificial Intelligence System (ALSAIS) was designed. Underlying ideas for this system originate from different existing approaches to concept-based learning [2, 3] and also from the attempt to mimic human infant learning in machine knowledge acquisition methods [4].

2. ALSAIS description

The main ALSAIS idea is based on emulation of infant mechanisms of knowledge acquisition (namely, concept-learning). Concept-learning requires splitting information into semantic and linguistic data channels. Both these channels are necessary for constructing concepts. Semantic data contains visual features and can be represented by information collected with a video camera. Linguistic data is information which includes speech features; it can be represented by an audio channel or a text input.

By combining simultaneous extracted audio and video features, dictionary database is constructed. This database contains pairs of lexical-semantic features. One such pair in this article is referred to as a lexical-semantic concept.

At first, new information is transmitted to the system's video and audio inputs. Semantic data is represented by video information of a scene; linguistic data is represented by audio information. Each grabbed video frame contains video description of the scene with some object, which visual features can be extracted from the frame. There is also a predefined condition that the objects of interest should be moving. Thus, the first-priority task is to find objects by performing motion segmentation. After that, the object is represented by a feature vector extracted from its image.

Although similar actions are performed with the audio channel, there is some difference. It is well-known that speech directed to infants is often redundant. The construction of an audio feature vector representing speech lexemes in this work is based on this assumption.

After construction of visual and audio feature vectors, their simultaneous analysis should be performed. Thus, the next task is to find correlations between audial and visual feature vectors. To do this, characteristics of the speech directed to infants should be taken into account. Here, we use mutual information as the learning criterion, which is quite popular in the domain of learnable computer vision systems [5], but first of all, feature vectors must be prepared for using. Our system performs automatic clustering in feature space and separate visual feature vectors between them. This gives us an opportunity to get particular features which describe parts of the object; we will refer to these features as visual cluster's characteristics.

By using the maximization of mutual information between the audial lexemes and the visual cluster's characteristics, the system constructs a dictionary of their pairs and calculates their estimated probabilities. The system only adds the pairs with maximum mutual information to our database.

ALSAIS can be described as a set of modules solving the following tasks: segmentation; constructing the visual lexemes feature vectors; constructing the audial lexemes feature vectors; clustering the feature vectors; maximization of mutual information.

All of these steps are repeated yielding an ability to add new concepts and their features to the knowledge base. Thus, the constructed knowledge base will contain all the descriptions of the objects in the scene. This knowledge allows the computer system to collect information about the objects being learnt.

2.1. Segmentation. As it was mentioned above, the process of segmentation consists in separating background pixels and object pixels in the grabbed image. In our system, two algorithms are combined together: optical flow and SURF descriptor.

2.2. Visual lexemes building. Each grabbed frame can be described by a visual feature vector. There are particular visual events which correspond to considerable changes in the current visual scene. The absence of the perceptible optical flow changes in the grabbed frame is used as a separator between the visual events. This idea helps us to group feature vectors belonging to the same object presented in different frames.

2.3. Audial lexemes building. The feature vectors of audial lexemes are extracted from the input audio channel. Lexemes can be obtained by the assumption of redundancy of speech directed to infants. For this purpose the algorithm of finding longest high-frequency substrings is used.

2.4. Data clustering and the maximization of mutual information. Clustering is used to group particular object features in clusters.

Mutual information shows the link between audial and visual lexemes. By calculating mutual information between sequences of visual and audial features, audial-visual pairs are constructed. Pairs with largest amount of mutual information surpassing a threshold are added into the concept database.

3. System learning results

ALSAIS unsupervised learning is evaluated by additional module – Control video evaluation. This module contains several types of experimental tests of learning. Three types of objects are examined in testing video data:

1. Objects, for which the system was trained, for example: comb.

2. Other exemplars of the same class, for which the system was trained, for example: other exemplars of comb.
3. Objects which are unknown to the system.

The particular results of the control video evaluation module are represented in table 1. Each category is marked by the number in brackets. The experiment consists in applying the final knowledge base to the test video, which contains different objects. In this experiment, only feature extraction and recognition were used. The extracted features were matched with the existing concepts in the database.

Table 1. Particular learning results for several objects.

Lexemes	Object							
	Staple (1)	Comb (1)	Camera (1)	Toy (1)	Timer (1)	Remote control (3)	Toy (2)	Comb (2)
Comb	335	2473	393	336	212	357	936	1628
Jpen	112	116	20	163	23	167	125	202
Pen	190	118	19	252	51	260	179	254
Photoapp	516	714	1624	414	273	231	547	504
Stepler	1859	1279	921	529	481	435	1552	1076
Tea	397	487	148	486	163	371	813	592
Timer	752	878	625	702	560	370	735	741
Toy	635	756	411	2388	279	1097	1001	1062

To simplify the result information formed by the control video evaluation module, the digits in table 1 are represented by a number of similar features between the lexemes and the object. Evaluations show that the system unsupervised learning process is successful in more than 90% of the presented objects.

Acknowledgements
This work was financially supported by the Ministry of Education and Science of the Russian Federation, and by Government of Russian Federation, Grant 074-U01.

References

[1] Potapov A S and Rozhkov A S 2013 Cognitive robotic system for learning of complex visual stimuli *Proc. AIP Conf.* **1537** 54–59

[2] Liu J and Bhanu B 2002 Learning Semantic Visual Concepts from Video *Proceedings International Conference on Pattern Recognition* **2** 1061–1064

[3] Jia Y, Abbot J, Austerweil J, Griffiths T and Darell T 2013 Visual Concept Learning: Combining Machine Vision and Bayesian Generalization on Concept Hierarchies *Advances in Neural Information Processing Systems* **26** 1842–1850

[4] Roy D and Pentland A 2002 Learning words from sights and sounds: a computational model *Cognitive Science* **26** 113–146

[5] Potapov A S, Malyshev I A, Puysha A E, Averkin A N 2010 New paradigm of learnable computer vision algorithms based on the representational MDL principle *Proc. SPIE* **7696** 769606

Evaluating the effectiveness of correlation digital speckle photography method for non-destructive testing of rough surfaces

A E Gorjunov[1], P V Pavlov[1], N V Petrov[2]
[1]Military Educational and Scientific Center Air Force "Air Force Academy named after prof. N.E. Zhukovsky and Y.A. Gagarin", 394064, Starih Bolshevikov 54-a, Voronezh, Russia
[2]Saint Petersburg National Research University of Information Technology, Mechanics and Optics, Kronverkskiy pr, 49, Saint Petersburg 197101, Russia

E-mail: fadedflower@mail.ru

Abstract. The work is devoted to the investigation of the process of reflection of laser beams from rough surfaces with different degrees of roughness. Based on the results of numerical experiments the applicability of the method of correlation analysis of speckle-structures for non-destructive inspection of surfaces with different parameters of roughness was evaluated. The optimal ratios between the parameters of the rough surfaces and parameters of the optical system, which provide the best efficiency of the method, were determined. It was established that in case of the increase of the number of topological charges the sensitivity of the optical system to changes of parameters of surface roughness increases.

1. Introduction

Nowadays, in connection with a wide variety of element base of optoelectronics the trend of the growth in production of devices for non-destructive control based on principles of non-contact methods of laser diagnostics is remarkable. One of the promising methods of non-destructive inspection is a digital speckle photography [1] based on correlation analysis of speckle-structures generated by reflection of coherent radiation from a rough surface. Since the structure of the speckle pattern contains full information about spatial roughness, this method can reliably distinguish surfaces with different parameters Ra.

To improve the efficiency of the method the use of vortex wave fronts as the probe radiation has been proposed [2], the mathematical model that provides the description of the spiral laser beams propagating in free space and their reflection from rough surfaces was also developed [3]. However, a detailed study that would identify the optimal parameters of the wave front (wavelength, the value of the topological charge) for the diagnosis of investigated surfaces which are characterized by different parameters (*Ra* and *Sm*) was not carried out. This work was undertaken in order to fill this gap. We present the results of numerical simulation of propagation and reflection of vortex and plain wave fronts from rough surfaces.

2. Numerical simulation

The simulation of propagation of wave fronts in free space (as well as their interaction with objects, characterized by different roughness parameters) is based on the equations of scalar diffraction theory [3]. The calculation of the wave front propagation from the diffraction optical element (DOE) plane (x', y') into the object plane (x, y), and from the object plane (x, y) into registration plane (X, Y) was performed by method of convolution of the original field with impulse response:

$$u(x,y,l) = FFT^{-1}\left\{FFT[u_0(x',y')] \times FFT\left[\frac{exp\{ikr\}}{i\lambda r}\cdot\frac{l}{r}\right]\right\} \quad (1)$$

Here $u_0(x', y')$ and $u(x, y, l)$ are functions that describe the spatial distribution of the complex amplitude of the wave field in the DOE plane and the object plane respectively; l is the distance between (x', y') and (x, y) planes; λ is the wavelength, $k = 2\pi/\lambda$ is wave vector; r is the distance from the center point of the (x', y') plane to each point in the (x, y) plane; FFT and FFT^{-1} are direct and inverse Fourier transforms. For the calculation of the wave front propagation from object plane (x, y) to the registration plane (X, Y), it is necessary to make the appropriate substitutions: l to L; $u_0(x', y')$ to $u(x, y, l)$; $u(x, y, l)$ to $U(X, Y, L)$.

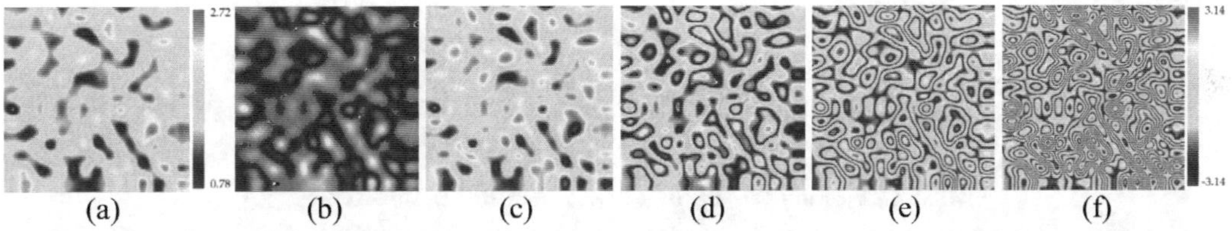

Figure 1. Spatial distribution of phase delay (in radian) of the flat wavefront (λ = 1 µm) in the plane of rough surface in case of reflection for different roughness, defined by value of maximum relief deviation h_{max}. h_{max}= 0,25 µm (a); 0,5 µm (b); 1 µm (c); 2 µm (d); 4 µm (e); 8 µm (f).

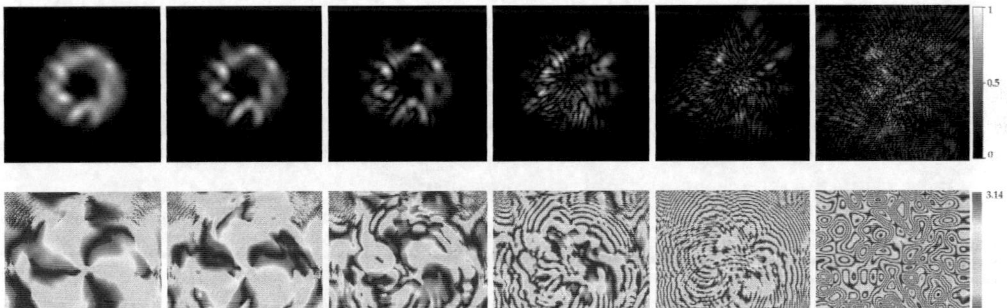

Figure 2. Spatial distribution of amplitude, in arbitrary units (top row) and phase (bottom row) of the vortex wavefront with topological charge $\rho = 3$ in the registration plane in case of reflection from rough surface with different values of maximum relief deviation h_{max}. h_{max}= 0,25 µm (a); 0,5 µm (b); 1 µm (c); 2 µm (d); 4 µm (e); 8 µm (f); λ = 1 µm.

As a result of the simulation (Figures 1, 2) the limit values of roughness *Ra* of the surface (for different values of the radiation wavelength), which are available for diagnosis by speckle photography using correlation analysis, were obtained. Limit values of maximum relief deviation h_{max}, which are available for diagnosis by a correlation analysis were: $h_{max} \approx 8\lambda$ in the case of reflection and $h_{max} \approx 32\lambda$ in the case of transmission schemes.

Obtained spatial intensity distributions in the registration plane were processed with correlation image processing software. For each intensity distribution obtained for different h_{max} values, correlation interval of the speckle pattern was calculated [2]. As a result, limit values of the maximum relief deviation h_{max}, which are available to diagnose by correlation speckle photography method, were established both for vortex and plain wavefronts. These values are presented in Table 1.

Table 1. Limit values of the h_{max} for schemes "reflection" and "transmission"

	reflection			transmission		
	Wavelength, μm					
	$\lambda = 0.5$	$\lambda = 1$	$\lambda = 1.5$	$\lambda = 0.5$	$\lambda = 1$	$\lambda = 1.5$
h_{max}, μm (vortex front)	3	7	9	15	32	48
h_{max}, μm (plain front)	4	8	11	16	34	50

In the case of using the vortex wavefront, the analysis of the influence of the value of topological charge ρ on the sensitivity of the diagnostic system was also carried out. Here, the reflection scheme was used. The obtained results are presented in the Table 2 for $\lambda = 0.5$μm.

Table 2. Values of the R_{cor} for scheme "reflection"

h_{max}, μm	R_{cor}, μm			
	flat wavefront	$\rho = 1$	$\rho = 2$	$\rho = 3$
0.2	14.8851	13.9839	11.7698	8.04026
1	12.4264	11.438	9.4959	7.24566
2	3.4224	3.078	2.62957	2.18205

As a conclusion, this paper describes the investigation of the limit values of the parameters of surface roughness that are available for diagnosis by digital speckle photography using correlation analysis with vortex and plane wave fronts. The assessment of the impact of the number of topological charges on the sensitivity of the diagnostic system is presented.

Acknowledgments
Nikolay Petrov acknowledge financial support from Russian Ministry of Education and Science project within the state mission for institutions of higher education (Agreement No 2014/190).

References

[1] Goodman J W 2006 *Speckle Phenomena in Optics: Theory and Applications* (Greenwood Village: Roberts and Company Publishers)
[2] Malov A N, Pavlov P V 2012 *Computer Optics* **36** 365
[3] Petrov V N , Pavlov P V , Malov A N 2013 *Quantum electron.* **43** 582

Matching of Aerospace Photographs with the use of Local Features

R Malashin[1]
[1]Saint Petersburg National Research University of Information Technologies, Mechanics and Optics, 49 Kronverkskiy pr., Saint Petersburg 197101, Russia

Email: malashinroman@mail.ru

Abstract. Matching of digital images is very challenging computer vision problem. The aim of investigation was developing of algorithms for matching real aerial and cosmic photographs. In the proposed methods, images are described locally by scale and rotation invariant descriptors. Reliability and high accuracy of the algorithms has been achieved by combining dense keypoint detector and robust descriptor with complex procedure of outlier elimination. The algorithms are capable of making correct decisions when number of local mismatches is more than 99%.

1. Introduction

The main difficulty in matching of aerospace images is that appearance of the Earth surface changes dramatically as the result of season, day-time, or viewpoint changes. In general case, object specific information also cannot be used due to wide diversity of possible content of these images. The target independent structural analysis is the only way for solving the task of matching of aerospace images [1], and there are two possible approaches for accomplishing it.

The first approach is to construct a very robust description of image that will capture only or almost only the invariant image features. The examples of successful implementation of this approach in the task of registration of aerospace images can be found in [2,3]. These methods are based on contour structural elements such as segments of straight lines, blobs and corners. These elements are not unique enough. That's why, plane structure of the Earth surface is assumed, and complex mutual relations of elements are necessarily taken into account. These methods cannot deal with non-planar surfaces.

The approach applied in this paper is based on matching of local features of images. Descriptors invariant to scale and rotation (like those applied in SIFT [4] and SURF [5]) are constructed in the regions of interest of images and then are matched to each other independently. Intuitively, our approach seems to be worthwhile of usage because some regions of aerospace images are almost insensitive to season and day-time changes. The local features extracted from such stable regions can be matched correctly, but another hard problem should be overcome. The great number of outliers arises from unstable regions and non-overlapped regions of images. Therefore the overall quality of matching depends significantly on the applied algorithms of eliminating outliers.

2. Proposed methods

Below there is a brief description of the proposed algorithms for feature detection, description and matching.

2.1. Detection and description of local features

Recently, there have been developed many different algorithms of keypoint detection and description. The experiments carried out with cosmic and aerial pictures have shown that the best results can be achieved using the SURF detector and the SIFT descriptor together.

The SIFT detector (as compared to the SURF detector) produces too many small keypoints that are less discriminative. The MSER [6] algorithm detects too few keypoints and can't provide needed "density" of description. The ORB [7] detector (and some other detectors) can't provide uniform distribution of keypoints in scale space.

The SIFT descriptor appeared to be more robust than other algorithms of local description. It is able to tolerate significant appearance changes of objects. The SURF descriptor captures too much object specific information and may be used when the alterability of territory appearance is not excessive. Binary descriptors (BRISK [8] and others) appeared to be inapplicable to aerospace images because they operate directly with pixel brightness (instead of intensity gradients) that is very sensitive to day-time and season changes.

2.2. Keypoint matching

Despite the appropriate selection of keypoint detector and descriptor (as it was discussed above), most of keypoint descriptors still are matched incorrectly. The outliers can be eliminated by means of a structural analysis that is described below.

2.2.1. Nearest Neighbor ratio versus 1-to-n matches exclusion. One of the most popular technique to eliminate outliers is use of nearest neighbor ratio as signal of mismatch [4]. Several experiments with manually labeled data were made. They showed that significant decrease of the number of inliers would result from thresholding ratio of distances to two closest neighbors in feature space when the images being matched are rather "difficult" for comparison. Figure 1 shows that most of inliers of "hard" image pairs refer to high ratios of distances to closest neighbors. It appeared that better results can be achieved by eliminating the ambiguity of matches that link several keypoints from the first image with one point of the second image. A single unambiguous match corresponding to the shortest distance in multidimensional feature space remains after such elimination. The n-to-1 matches have higher probability to be incorrect, and moreover, they have higher probability to form incorrect (false) clusters of matches with relevant parameters (Figure 2). The algorithms of cluster analysis applied for elimination of wrong matches are described below.

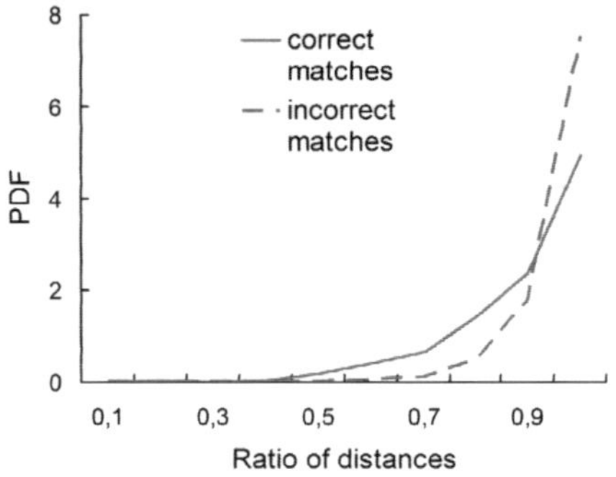

Figure 1. Probability density function of correct and incorrect matches. Generated from manually labelled data.

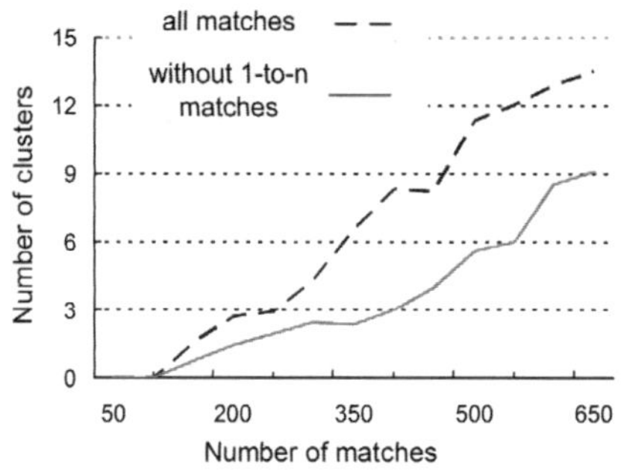

Figure 2. Average number of false clusters of matches (with 3 and more votes) before and after exclusion of 1-to-n matches. The graph was generated by matching hundreds of images having non-overlapping content.

2.2.2. Geometric constraints. Many mismatches can be eliminated by fitting a planar model to data generated by matching of descriptors. As soon as the Earth surface is not planar in general case, significant deviations from planar model should be accepted. Similarity transform model appeared to be more robust (with respect to outliers and "non-planar data") than more general affine transform model. Preliminarily, rough parameters of similarity transform are estimated by Hough Clustering of parameters of local co-ordinate transforms that are described by individual matches [4]. The bin size of Hough accumulator smaller than in [4] is used because the nearly planar scenes are considered. By the same reason, only the biggest cluster in Hough accumulator is further analyzed. The RANSAC algorithm is then used to estimate the parameters of similarity transform precisely and eliminate some more outliers from cluster. The mutual position, scale and orientation of matched keypoints are taken into account when fitting the model. Finally, the parameters of mutual affine transform of co-ordinate systems of two matched images are estimated by least square solution [4]. If the number of matches that passed all listed above tests falls below predefined threshold (with respect to initial number of matches) two photographs are considered to differ too strong from each other.

3. Experimental results

The proposed algorithms where successfully tested based on a series of aerial and cosmic photographs with strong appearance changes caused by season, day-time and viewpoint variation. The anthropogenous (manmade) changes were also considered by matching old photographs of city landscapes with modern ones. And moreover, SAR images were successfully matched with ones formed by optical sensors. The reliability and accuracy of the algorithms are achieved by eliminating of huge number of outliers by the use of geometrical constraints. The correct solutions can be made even based on small correct subsamples of the whole initial multitude of point-to-point matches (in some cases based on less than 0.1% of the whole multitude of matches).

Acknowledgements
This work was supported by the Ministry of Education and Science of the Russian Federation, and by the Government of the Russian Federation, Grant 074-U01.

References

[1] Lutsiv V, Malyshev I, Pepelka V and Potapov A 2002 *Proc. of SPIE* **4741** 351–62
[2] Lutsiv V., Malyshev I and Potapov A 2003 *Proc. of SPIE* **5238** 164–75
[3] Lutsiv V, Malyshev I and Pepelka V 2001. *Proc. of SPIE* **4380** 174–83
[4] Lowe D 2004 *Int. J. of Computer Vision* **60** 91-110
[5] Bay H, Tuytelaars T and Van Gool L 2006 SURF: Speeded Up Robust Features *Proc. 9th European Conf. on Computer Vision (Graz, Austria, May 7-13)* pp 404–417
[6] Matas J, Chum O, Urban M and Pajdla T 2002 Robust wide baseline stereo from maximally stable extremal regions *Proc. of British Machine Vision Conf (Cardiff, UK, September 2-5)* pp 384–396
[7] Rublee E, Rabaud V, Konolige K and Bradski G 2011 ORB: An efficient alternative to SIFT or SURF *Proc. of Int. Conf. on Computer Vision (Barcelona, Spain, November 8-11)* pp 2564–2571
[8] Leutenegger S, Chli M and Siegwart R 2011 BRISK: Binary Robust Invariant Scalable Keypoints *Proc. of Int. Conf. on Computer Vision (Barcelona, Spain, November 8-11)* pp 2548–2555

Determining the orientation of the observed object in three-dimensional space using stereo vision methods

S Ponomarev[1]
[1]Saint Petersburg National Research University of Information Technologies, Mechanics and Optics, 49 Kronverkskiy pr., Saint Petersburg 197101, Russia

E-mail: sv.v.ponomarev@gmail.com

Abstract. The task of matching image of an object with its template is central for many optoelectronic systems. Solution of the matching problem in three-dimensional space in contrast to the structural alignment in the image plane allows using a larger amount of information about the object for determining its orientation, which may increase the probability of correct matching. In the case of stereo vision methods for constructing a three-dimensional image of the object, it becomes possible to achieve invariance w.r.t. background and distance to the observed object. Only three of the orientation angle of the object relative to the camera are uncertain and require measurements. This paper proposes a method for determining the orientation angles of the observed object in three-dimensional space, which is based on the processing of stereo image sequences. Disparity map segmentation method that allows one to ensure the invariance of the background is presented. Quantitative estimates of the effectiveness of the proposed method are presented and discussed.

1. Introduction

The fundamental problem arising in the process of matching is the problem of prior uncertainty due to the variability of object and shooting conditions (angle and lighting), as well as a lack of information about the properties of object. There are two basic approaches to the determination of the angle of observation or combination of reference and current visual information that can be divided by the fact whether there is alignment in the image plane or in three-dimensional space. In the first case, the current image should be aligned with the template image formed for the corresponding angle, distance, lighting conditions and background [1]. In this approach, the number of templates for each object can be quite large, and the process of aligning with each of them to establish the angle could become resource consuming.

In the second case, the alignment should be made for the reconstructed disparity map and three-dimensional model of the object [2]. Data on the three-dimensional shape of an object is initially presented in a form invariant to lighting conditions. Invariance w.r.t. background is also easily reached by depth map segmentation as a background has zero disparity. Furthermore, invariance w.r.t. distance to the object which is extracted by stereovision methods is automatically achieved. These features of the second approach can provide a higher probability of correct matching. To determine the relative orientation of the observed object in the processing sequence of stereoscopic images one can use the methods of alignment the three-dimensional model of the object reconstructed on the basis of dense depth maps for images from a sequence of stereo pairs. In this paper, we propose a method that uses the approach described above to determine the orientation angles of the object in three-dimensional space.

2. Modeling and obtaining baseline data

In the course of this work orbital space flight simulator Orbiter 2010 [3] was used as a basis for the modeling system. Software that uses a graphical module OGLA, which allows visualization of the processes of change the relative position of the observer and the investigated object, was developed. This program implements emulation for a stereo imaging of observed scene with the automatic change of camera parameters that allows the study of algorithms in various stereovision conditions.

At the second stage additive white Gaussian noise with the specified signal/noise ratio is superimposed on the obtained images of stereo pair. The resulting disparity maps were generated using Semi-Global Stereo Matching algorithm (SGM) [4]. This algorithm is showed its effectiveness at the stage of preliminary studies, because firstly, it was comparable in speed with local algorithms, and secondly, as compared with Graph cuts and Belief Propagation, it provides a better smoothness for various disparity values.

3. Disparity map segmentation algorithm

Due to the fact that even at the signal / noise ratio of 21, there is considerable distortion in the background region of the resulting disparity map, an important part of the work was to develop a method of disparity maps filtration in order to separate the observed object and background. Suggested algorithm consists of combination of SURF for key points detection and binarization with the addition of morphological operations. It can be represented as the following steps, as shown in Figure 1.

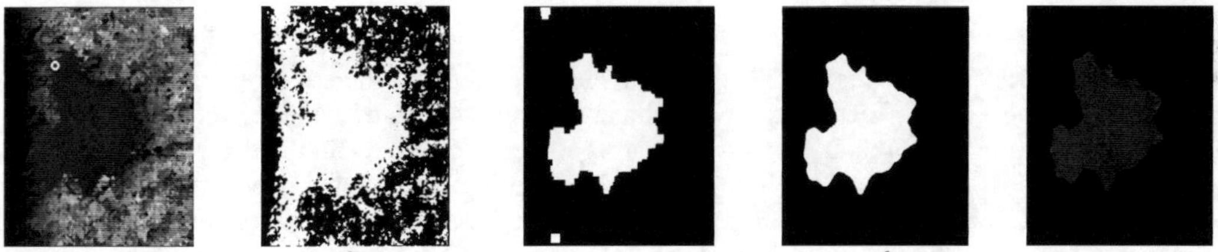

Figure 1. Steps of filtering disparity maps: smoothed disparity map with highlighted key point (a), application of thresholding (b), application of opening (c), median filtering and selection of the largest contour (d), disparity map after filtering and segmentation (e).

4. Algorithm for determining the orientation angles of the observed object

4.1. Peculiarities of the algorithm. The input data of the developed algorithm is a pair of segmented disparity maps. At the first stage with the knowing of the virtual camera of the simulation system internal parameters the reconstruction of point clouds in three-dimensional space is carried out. Then in the case of temporary delays or low frequency mode rough alignment on the 4 corresponding points is performed using previously obtained key points (SURF). To obtain more accurate values of displacements iterative optimization is performed using the ICP method [5].

The result of ICP algorithm is a rigid transformation of initial point clouds, which may be represented by the transformation matrix. Knowing the transformation matrix the Euler angles can be restored. Excluding shift angles it can be expressed for the three axes of rotation of the elements of the transformation matrix as follows:

$$\boldsymbol{\theta}_x = \operatorname{atan}2(a_{32}, a_{33}) \tag{1}$$

$$\boldsymbol{\theta}_y = \operatorname{atan}2(-a_{31}, \sqrt{a_{32}^2 + a_{33}^2}) \tag{2}$$

$$\boldsymbol{\theta}_z = \operatorname{atan}2(a_{21}, a_{11}) \tag{3}$$

4.2. Experimental results. The proposed algorithm has been tested for the following situations:

1. Object of simple shape «Pioneer-1», turn on the Y axis by 10 degrees.
2. Object of simple shape «Carina», turning on the X axis by 15 degrees and the Y-axis by 10 degrees.
3. Object of complex shape «Dragonfly», turning on the Y-axis by 20 degrees.

Table 1. Quantitative evaluation of the effectiveness of the algorithm.

Situation	Axis	Initial angles, degrees	Obtained angles, degrees	Error of the algorithm, degrees.
	OX	0	-0,43	0,43
1	OY	10	7,87	2,13
	OZ	0	0,45	0,45
	OX	0	1,82	1,82
2	OY	10	6,33	3,67
	OZ	15	13,25	1,75
	OX	0	0,16	0,16
3	OY	20	15,16	4,84
	OZ	0	2,73	2,73

5. Conclusion

Analysis of the experimental results of this algorithm allows saying that there is a relationship between the magnitude of the error in determining an object's orientation and the direction of change in orientation. The rotation along the OY axis significantly increases the error in the determination of the angle, which can be explained by a small change of the resulting maps range due to the symmetrical shape of the observed objects. At the same time, rotations in other directions determined with higher precision, the value of the error amounted to 1-2 degrees. The developed algorithm allowed demonstrating the principal possibility of determining orientation of the observed object using stereo vision methods, but for more detailed qualitative and quantitative evaluation of the algorithm one should develop more detailed standards of the object.

Acknowledgements

This work was supported by the Ministry of Education and Science of the Russian Federation, and by Government of Russian Federation, Grant 074-U01.

References

[1] Potapov A S, Luciv V R 2003 Information-theoretic approach to image description and interpretation *Proceedings of SPIE* **5400** 277–283

[2] Arun K S, Huang T S, Blostein S D 1987 Least-squares fitting of two 3D point sets *IEEE TPAMI* **9** 698-700

[3] Paton M D and McSorley A 2009 Virtual prototyping of a lander for the fast transfer of humans to Mars *Journal of the British Interplanetary Society* **62** 369-377

[4] Hirschmuller H 2008 *IEEE TPAMI* **30** 328–341

[5] Chen Y, Medioni G 1991 Object Modeling by Registration of Multiple Range Images *IEEE ICRA* 2724–2729

Image inpainting based on stacked autoencoders

O Shcherbakov[1] and V Batishcheva[2]
[1]Saint Petersburg National Research University of Information Technologies, Mechanics and Optics, 49 Kronverkskiy pr., Saint Petersburg 197101, Russia
[2]Saint-Petersburg State University, 7-9 Universitetskaya nab., St.Petersburg 199034, Russia

E-mail: scherbakovolegdk@yandex.ru, elokkuu@gmail.com

Abstract. Recently we have proposed the algorithm for the problem of image inpaiting (filling in occluded or damaged parts of images). This algorithm was based on the criterion spectrum entropy and showed promising results despite of using hand-crafted representation of images. In this paper, we present a method for solving image inpaiting task based on learning some image representation. Some results are shown to illustrate quality of image reconstruction.

1. Introduction

Image inpainting problem (filling in missing or damaged regions in images) presenting its own theoretical interest has also wide field of applications. Such applications could be, for example, image recovery after lossy image transmission or communication and different kinds of painting restorations. Some partial solutions for this task have already been developed. There are methods based on wavelet representation of images [1], variational analysis [2], structural features [3], interpolation models, algorithmic probability (ALP) [4] and etc. Mostly, these methods use particular hand-crafted image representation. However under ALP approach the image inpainting task has been stated for generalized image representation [4], and similar approach has been investigated for representation learning [5]-[6], but not in the context of image inpainting. Thus the problem of learning representations is of interest here.

One of the approaches to automatic feature learning from unlabeled data is stacked autoencoder (AE) that presents a special type of artificial neural networks (ANN). This approach is interesting for image inpainting task because architecture of AE allows not only encoding some features in its hidden units but also reconstructing an input pattern when decode mode of AE is on. However, this feature of AE does not allow one to solve image inpainting task directly. Nevertheless, this approach to feature learning can be extended to perform image inpainting.

2. Autoencoders for learning representation

Autoencoders present the unsupervised learning algorithm that sets output values to be equal to the inputs. In other words, AE tries to learn approximation to identity function so as output vector is going to be similar to input vector. Such identity function demands some constraints on network structure. One of these constraints is to limit the number of hidden units which forces network to learn compressed representation of the input.

Thereby AE has to reconstruct full input vector by means of given hidden unit activations. If input vector values were completely random, then compression and reconstruction tasks would not be easy. However, if there are some regularities in the input data (e.g. some features of patterns are correlated), then this network will be able to discover some of these regularities. Such approach seems very similar to principal component analysis (PCA), but it already can partially help in solving inpainting problem.

In this work we use stacked AEs which provide architecture for learning deep representations. In Figure 1 b) one can see that every hidden layer of the network except the last one might be considered as input vector of the next AE. This structure allows the stacked AE to learn more abstract features and to train each AE one after another. Also this feature makes the stacked AE less trivial comparing to PCA. Training the stacked AE is reduced to minimization of simple linear form for each simple AE.

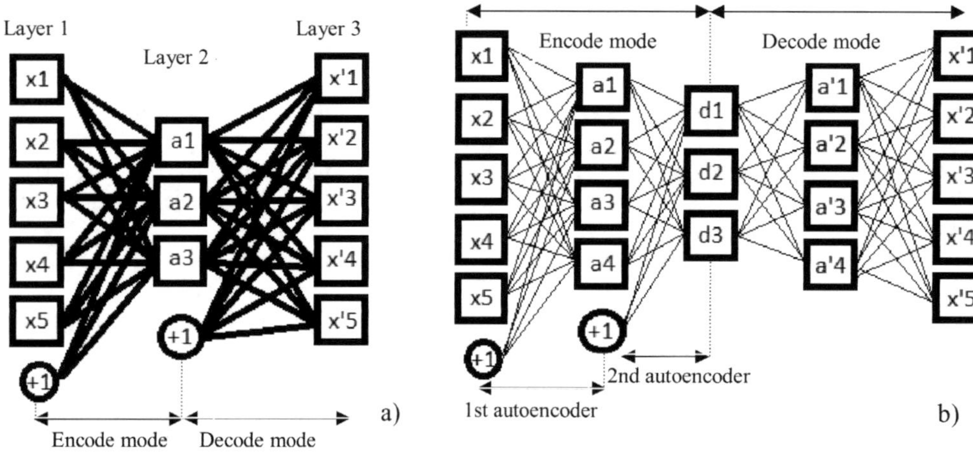

Figure 1. AE (a) and stacked AE (b) structure. Figure (a) shows simple AE structure that works similarly to PCA. Figure (b) shows multi-layer structure of stacked AE which is composed of two simple AEs.

3. Proposed algorithms

In this paper we propose the next two algorithms for image inpainting based on stacked autoencoders. The first one represents the iterative procedure which can be described by several steps:

1. Stacked AE are trained using training set of images
2. Corrupted image (e.g. image with occluded region) is passed to the input of the trained stacked AE
3. Network's response to the input stimulus (reconstructed image) is calculated
4. Corrupted areas of the input image are replaced by reconstructed ones
5. The image with replaced areas becomes the input vector of the stacked AE
6. Steps 3-5 are repeated until stopping criterion (e.g. number of iterations) is met

The second algorithm is based on the assumption that reconstructed region of the image can be described as fixed point (also known as fixpoint or invariant point) of some very complex and multidimensional function. Therefore, that function may have more than one such fixed point. The idea is to find optimal fixpoint using metaheuristic search such as genetic algorithms (GA). Fitness function for each point individual can be calculated as standard deviation (SD) between input and reconstructed by stacked AE images.

4. Experimental results

For experimental evaluation we used handwritten digits database and FEI University face database. Stacked AE was trained on the training set consisting of different digits or different faces. Test images were created by cutting some region of interest from an image. Image inpainting algorithms were tested with different parameters such as number of images in training set, number of hidden layers in AE, sizes of hidden layers, learning rate, number of training epochs and etc. Figure 2 shows some results of inpainting for digits and figure 3 shows inpainting results for faces.

a) b) c) d)

Figure 2. Inpainting results for handwritten digits images. (a) original image. (b) iterative algorithm inpainting result. (c) GA inpainting result. (d) corrupted image.

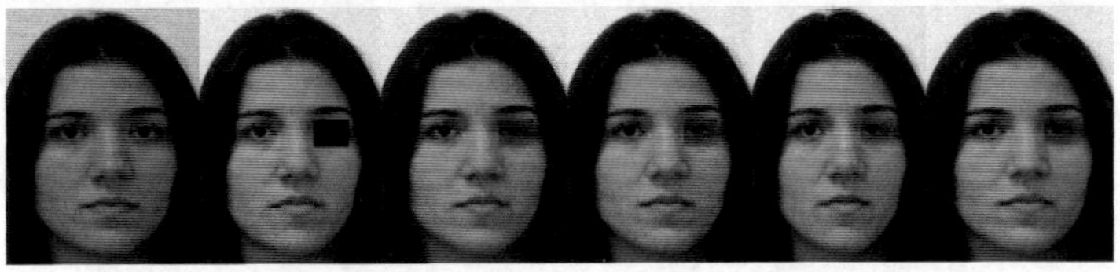

a) b) c) d) e) f)

Figure 3. Inpainting results for frontal facial images. (a) original image. (b) corrupted image. (c) iterative algorithm inpainting result with 3 hidden layers in AE. (d) GA inpainting result with 3 hidden layers in AE. (e) iterative algorithm inpainting result with 5 hidden layers in AE. (f) GA inpainting result with 5 hidden layers in AE.

5. Conclusion

We have considered the approach to image inpainting based on stacked autoencoders. This approach was implemented in two image inpainting algorithms and tested on two types of images handwritten digits images and frontal facial images. The results show that inpainting using GA approach has slight advantage compared to iterative inpainting procedure. Also increasing amount of hidden layers in stacked AE allowed gaining different kinds of inpainting results which might be caused by more abstract features learnt by the network on deeper layers.

Acknowledgements

This work was supported by the Russian Federation President's grant Council (MD-1072.2013.9) and the Ministry of Education and Science of the Russian Federation.

References

[1] Chan T. F., Shen J., Zhou H.-M. A 2006 Total Variation Wavelet Inpainting Model with Multilevel Fitting Parameters *Proc. SPIE 6313, Advanced Signal Processing Algorithms, Architectures, and Implementations XVI.*

[2] Arias P., Caselles V., Sapiro G. A 2009 Variational Framework for Non-local Image Inpainting *Proc. EMMCVPR'09*. 345–358.

[3] Shibata T., Iketani A., Senda Sh. 2011 Fast and Structure-preserving Inpainting Based on Probabilistic Structure Estimation MVA2011 *IAPR Conference on Machine Vision Applications, Nara, JAPAN* 22–25.

[4] Potapov A., Scherbakov O., Zhdanov I. 2013 Practical algorithmic probability: an image inpainting example *Proc. SPIE 9067, Sixth International Conference on Machine Vision.*

[5] Potapov A. S. 2012 Using multiple video-information representations in automatic image-analysis systems *Journal of Optical Technology* 79 689–692.

[6] Potapov A. S., Malyshev I. A., Puysha A. E., Averkin A. N. 2010 New paradigm of learnable computer vision algorithms based on the representational MDL principle *Proc. SPIE* vol 7696, 769606.

Use of spectral characteristics of DSLR cameras with Bayer filter sensors

P A Cheremkhin[1], V V Lesnichii[2,3], N V Petrov[2]
[1] National Research Nuclear University MEPhI (Moscow Engineering Physics Institute), Kashirskoe shosse 31, Moscow 115409, Russia
[2] Saint Petersburg National Research University of Information Technologies, Mechanics and Optics, 49 Kronverkskiy pr., St. Petersburg 197101, Russia
[3] University of Freiburg, Freiburg P.O. Box 79085, Germany

E-mail: cheremhinpavel@mail.ru

Abstract. The expensive photosensors of scientific cameras are commonly used in the wide variety of research fields. However, photosensors that are implemented in DSLR cameras are seen to be an appropriate substitution in order to decrease price/quality ratio or even receive additional features. In this article different scientific applications of DSLR cameras' photosensors with Bayer filter as well as calibration methods of its spectral characteristics are discussed. The approach, based on determination of latter and usage of its features, is shown to increase SNR of the color reconstructed images in digital holography.

1. Introduction

Nowadays, the two most common types of photosensors of digital single-lens reflection (DSLR) cameras that are used: CMOS and CCD. As it was shown, for example, in work [1], such cameras (as Canon EOS 400D) have characteristics that do not yield up to the ones that specific sensors destined for scientific research have. Nevertheless, a wide usage of DSLR cameras for photometry is impeded. As we can see, there are two main reasons for this:

- A conviction, that the loss of information about registered intensity distribution caused by built-in cameras' software is inevitable;
- An absence of comprehensive radio- and photometric data (characteristic curves, noise and spectral characteristics) for photosensors in public access.

In order to circumvent postprocessing, as, for example, gamma correction, one should perform a registration in RAW format. Then, the applying of a specific freeware such as dcraw [2] leads to the receiving of the linearized data directly from the camera's ADC without color interpolation and other possible unwanted postprocessing [3].

The determination of noise characteristics and linearity of the camera's response is also possible by means of dcraw. Such a calibration of cameras' photosensors and publication of these specifications (like in database [4] or in article [5]) promotes popularization of DSLR cameras' usage for laboratory photometrical measurements.

Of all varieties of cameras' photometric values in this paper we focus on the RGB-sensor's spectral response, its analysis and usage.

2. Applications of spectral characteristics of DSLR cameras with Bayer filter and calibration methods

Spectral characteristics of sensors are demanded in a variety of scientific research fields, the most topical of which is the computational color constancy [6]. Other areas are machine vision, remote sensing of terrestrial surfaces [7], digital photography [8] and holography [9], and phase retrieval [10]. There are various methods of spectral calibration and characterization of sensors of digital cameras: by measuring of set of samples of known spectral reflections [11], by parametric fitting of sensor responses [12], hybrid method [13] and etc. EMVA offered a standard 1288 for estimation of integral spectral response of sensors [14], but it states that measurements in different color channels are necessary only in case of the sensor's quantum efficiency evaluation.

In works [6, 15] the computational method based on the standard camera model is proposed. The main advantage of this method is an opportunity of retrieving the function, which both compensates the gamma correction, vignetting and other possible nonlinear effects and linearizes the camera spectral response in all three color channels. The experimental load is similar to one used in works [16-17]. This method, however, leaves unexposed the opportunity of usage a weak sensor's response at wavelengths away from the peak of spectral sensitivity, which is considered, for example, in [17]. In the latter article a method of the spectral response estimation of DSLR camera's sensor separately over Bayer filter (Figure 1) is proposed. Using dcraw and simple software processing this method allows direct measurement of spectral characteristics for the whole visible range of wavelengths. In mentioned methods three spectral responses (for pixels of separate Bayer filter's channels: red, green and blue) are considered as an outcome.

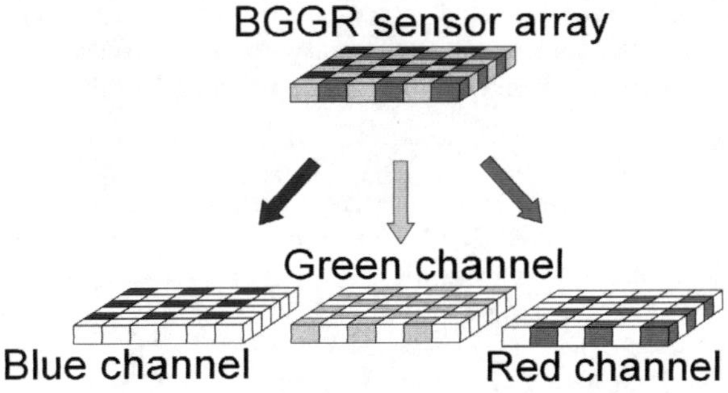

Figure 1. Colour channels of Bayer filter

3. Application of spectral characteristics of DSLR cameras' CCD and CMOS-sensors separately over the Bayer filter to digital holography

In [17] two applications of sensor's RGB spectral characteristics for issues of the digital holography were proposed. In this work another one is under consideration. It implies registration of three digital holograms at one camera shot. The below outlined sequence allows receiving of color reconstructed images of 3D scene with better SNR (signal-to-noise ratio) compared to obtained in [17]:

1. Choose three wavelength, spectral responses at which for complementary (to the chosen one) color channels are minimal;
2. Register simultaneously three digital holograms (or three diffraction speckle patterns, like it was done in [10]) at denoted wavelengths;
3. Digitally reconstruct images, interpolate them and receive color 3D image of the object.

Thus, using this approach, the benchmark data is not the spectrum of the light source, as it was in works [17] and [9] (coherent and incoherent source, respectively), but the spectral response of sensor.

As an example one can consider previously measured [18] responses of the Canon EOS 1000D, see Figure 2, where chosen according to the 1st step of the above mentioned method wavelengths are depicted by three color lines.

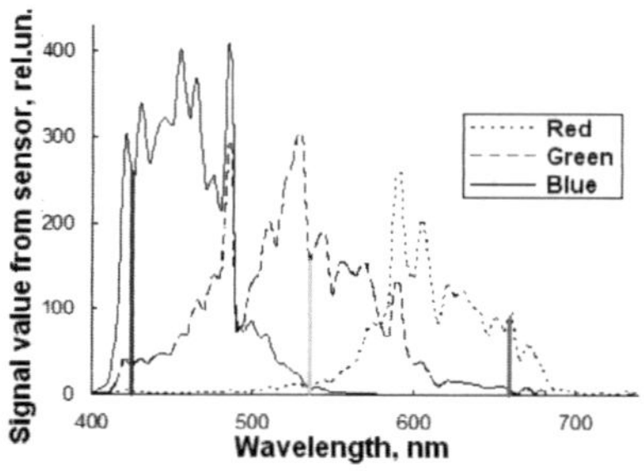

Figure 2. Spectral responses of color channels of sensor of camera Canon EOS 400D and three chosen wavelengths for the color digital hologram recording

Usage of such approach allows decreasing of image's noisiness by up to one order of magnitude because of the reduction of influence of complementary color channels illumination. Therefore the quality of reconstructed from digital holograms images, as follows from [18], is 10 times enhanced.

4. Conclusion

DSLR cameras can be successfully applied to different photometric tasks in case of RAW format registration and following usage of dcraw converter, since the retrieved response in this case is linear. The range of tasks, for solution of which the spectral response of the DSLR cameras' sensor is of interest, was outlined in current work. The analysis of spectral calibration methods was presented as well. The approach based on usage of the spectral characteristics' features was proposed. It provides a substantial enhancement of SNR of color images reconstructed from digital holograms.

Acknowledgments

Vasilii Lesnichii acknowledges the financial support from Government of Russian Federation, Grant No 074-U01. Nikolay Petrov acknowledges financial support from Russian Ministry of Education and Science project within the state mission for institutions of higher education (Agreement No 2014/190).

References

[1] Evtikhiev N N, Starikov S N, Cheryomkhin P A and Krasnov V V 2012 *Proc. SPIE, Intelligent Robots and Computer Vision XXIX: Algorithms and Techniques* vol 8301 (SPIE) p 830113
[2] Coffin D 2014 Decoding raw digital photos in Linux, http://www.cybercom.net/~dcoffin/dcraw
[3] Konnik M V, Manykin E A and Starikov S N 2010 *Quant. Electron.* **40** 314-20
[4] Camera Sensor Database, http://www.dxomark.com/Cameras
[5] Manders C and Mann S 2005 *Proc. ICICS (Bangkok)* (IEEE) p 1292-6
[6] Barnard K and Funt B 2002 *COLOR research and application* **27** 152-63
[7] Lebourgeois V, Bégué A, Labbé S, Mallavan B, Prévot L and Roux B 2008 *Sensors* **8** 7300-22
[8] Stevens M, PARraga C, Cuthill I, Partridge J and Troscianko T 2007 *Biol J Linn Soc* **90** 211-37
[9] Kim M K 2013 *Opt. Express* **21** 9636-42
[10] Petrov N V, Bespalov V G and Gorodetsky A A 2010 *Proc. SPIE* vol 7387 (SPIE) p 73871
[11] Hardeberg J Y, Brettel H and Schmitt F J M 1998 *Proc. SPIE* vol 3409 (SPIE) p 100-9
[12] Thomson M and Westland S 2001 *Color Research & Application* **26** 442-9
[13] Zhang H, Quan S and Liu H 2010 *Proc. Int. Congress on Imaging Science (Beijing)*
[14] http://www.emva.org/cms/upload/Standards/Stadard_1288/EMVA1288-3.0.pdf
[15] Barnard K and Funt B 1999 *Proc. SPIE, Human Vision and Electronic Imaging IV* vol 3644
[16] Vora P L, Farrell J E, Tietz J D and Brainard D H http://color.psych.ucsb.edu/hyperspectral/
[17] Lesnichii V V, Petrov N V and Cheremkhin P A 2013 *Opt. Spectrosc.* **115** 557-66
[18] Evtikhiev N N, Starikov S N and Cheremkhin P A 2013 *J. Opt. Technol.* **80** 301-8

Estimate of temperature stability of volume polymer holograms

P Kudriavcev[1], N Andreeva[1], O Andreeva[1]
[1] Saint Petersburg National Research University of Information Technologies, Mechanics and Optics, Kronverkskiy 49, St. Petersburg 197101, Russia

E-mail: Pasha-kudriavcev@mail.ru

Abstract. The results of experiment on the influence of temperature of the environment on the parameters of the volume polymer holograms are presented. The experiments were carried out on specialized holographic setup using digital procedures of recording information and data processing in the range of temperature change $(22\div32)^0$C. It was shown that for hologram-gratings with the thickness about of one millimeter with spatial frequency 350mm^{-1}, the relative change of the angle of diffraction during temperature changing in given range is $1{,}4\cdot10^{-6}$ K^{-1}.

1. Introduction

The recording media on the polymer base are widely used for recording information, including recording of holograms, holographic optical elements (HOE) and optical elements for systems of archival optical-holographic memory [1]. Volume media with physical thickness of the order of millimeter, intended for these purposes, must ensure the permanence of the registered structure in the process of its production and exploitation. However, the change of medium temperature leads to a change of the geometrical sizes of such structure and its optical constants.

The changes of the sizes of the medium and its average refractive index, as a rule, are of the same order of magnitude, which makes it difficult to estimate them, especially in cases when these changes are differently directed.

In this work the results of experiments that allow us to estimate the impact of temperature changes of the environment on the parameters of volume hologram-gratings are presented. The use of holographic procedures allows to diagnose negligibly small changes of parameters of the samples, which are due to temperature instability that usually unattainable by other methods of research.

2. Object of study

Samples of polymer recording medium "Difphen" developed for recording volume holograms were used as the object of study [1]. Samples of the material "Difphen" represent one of the modifications of polymer light-sensitive medium on the basis of phenanthrenequinone (PQ) and methylmethacrylate (PMMA), in which is used the principle of diffusion increasing of recorded information.

Changes in the samples during the temperature changes can be characterized by the coefficient of linear expansion (dl/dT) and the temperature change of the refractive index (dn/dT) of medium. These coefficients for media based on polymer PMMA are given in table 1.

Table 1. The temperature coefficients of linear expansion (dl/dT) and the changes of the refractive index (dn/dT) of media based on polymer PMMA [2].

Characteristic	Polymer, PMMA
Linear thermal expansion, dl/dT, K^{-1}	$(3.6 \div 6.5)\ 10^{-5}$
Refractive index, dn/dT, K^{-1}	$-\ 1.05 \cdot 10^{-4}$

When heating polymer sample increasing of its linear dimensions and the reduction the average refractive index happens, this leads to the change of conditions of reading holograms defined by the Bragg's ratio:

$$2D\sin\theta_0 = \lambda/n_0, \quad (1)$$

where D – period of hologram, λ – wavelength of radiation (in the air), θ_0 – angle of diffraction in the medium; n_0 – average refractive index of the medium. The studied holograms were recorded on spatial frequency $\nu = 330\ mm^{-1}$ ($D = 1/\nu = 3\mu m$).
In this work the changes of hologram were diagnosed by changing the angle of diffraction of radiation $\delta\theta$.

3. Procedure of experiments

For estimation of change of the angle of diffraction of radiation during reading holograms in different conditions were used two different procedures. In both procedures the sample with recorded hologram was heated in the temperature range of 20-40^{0}C with the same procedure. The heating was produced by a stream of warm air in order to avoid a sharp rise of temperature of the sample heater was switched on for 30 seconds with an interval of 60 seconds. Measurements of parameters of the hologram were carried out before heating of the sample and in the process of its cooling.

Procedure 1. The studied holograms was illuminated by divergent beam of radiation with the wavelength of λ=633 nm (He-Ne laser). The intensity distribution in the beam of the diffracted radiation was registered on CMOS - matrix of the camera (fig.1a). This allows obtaining information about contour of angular selectivity of the hologram and change of the angle of diffraction during single measurement act $\delta\theta$ [2]. The angular resolution of the matrix in this scheme is 0,02 mrad.

Procedure 2. The sample with the studied hologram is installed in the scheme in which this hologram was recorded and is illuminated by two coherent beams I_1 and I_2 which were used for recording holograms (fig.1b). During changing hologram parameters by changing temperature in the space of overlapping beams I_{1d} и I_{20}, I_{2d} и I_{10} the low-frequency interference pattern with period d was formed, which could be observed in the air during installing the screen or CMOS-matrix (fig.1c). In accordance with Bragg's condition and considering that the angle φ is small the period of interference pattern can be calculated by the formula $d = \lambda\backslash\varphi$, in our scheme $\varphi \equiv \delta\theta$.

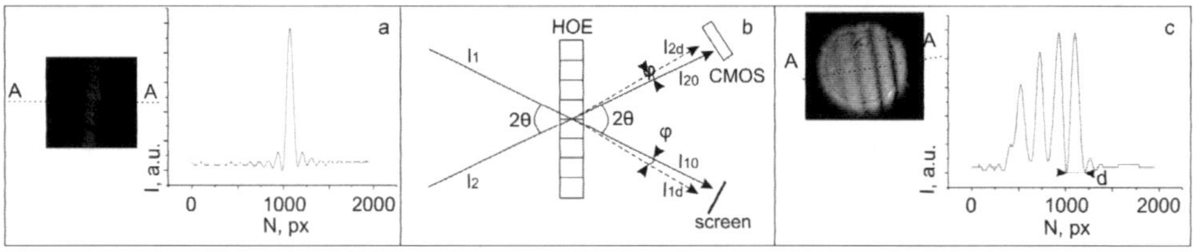

Figure 1. a - Procedure 1: distribution of radiation intensity in diffracted beam registered on the CMOS-matrix (left) and after processing (right); b, c - Procedure 2: scheme of the experiment (b); low-frequency interference pattern (c), registered on CMOS-matrix (left) and after processing (right). A-A – the line of scanning the results at digital processing.

4. Results of experiments and their analysis

The results of experimental measurements in the range of temperature change of the sample (22÷32)°C are shown on fig 2. Changing of position of the maximum of circuit of angular selectivity of hologram on CMOS-matrix (procedure 1) is given on fig.2a. It should be noted that measurements by different procedures were carried out by using radiation with different wavelengths (488 nm and 633 nm). Therefore, the comparison of changes of angle of diffraction should not be in absolute units, but in relative units, as it is done in table 2, line 4. Thus, relative change of angle of diffraction of polymer hologram in the visible spectrum during temperature change at one degree presents the value $\delta\theta/2\theta = (1{,}39\pm0{,}2)10^{-6}$.

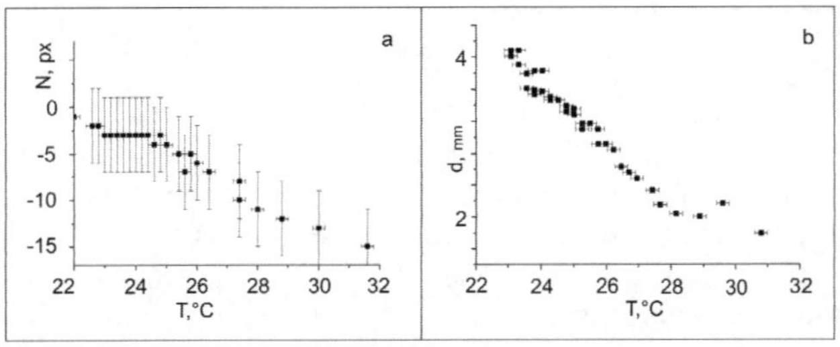

Figure 2. The results of experiments on influence of temperature changes on the hologram parameters: a - the shift of the maximum circuit of angular selectivity (procedure 1); b - change of the period of low-frequency interference pattern (procedure 2).

Table 2. The change of the angle of diffraction of radiation δθ* during reading volume polymer hologram-grating at change of temperature (T).

Number of measurement result	Range of temperature changes, ΔT	δΔθ	
		Procedure 1, λ = 633 nm	Procedure 2, λ = 488 nm
	1	2	3
1	(22 ÷31)°C	(0,30 ÷ 0,04) mrad	
2	(23 ÷31)°C		(0,19 ÷ 0,02) mrad
3	1K	(0,033 ÷ 0,004) mrad	(0,024 ÷ 0,002) mrad
4	1K	$\delta\theta/2\theta = (1{,}41\pm0{,}15)10^{-6}$	$\delta\theta/2\theta = (1{,}37\pm0{,}14)10^{-6}$
5	1K	$\delta\theta/2\theta = (1{,}39\pm0{,}2)10^{-6}$	

5. Conclusion

Measurements characterizing temperature stability of volume polymer holograms in the range of temperature change (22÷32)°C were carried out. Two different methods revealed the presence of changes of the angle of diffraction of radiation during temperature changes. It was found that the increase of temperature leads to increase of the angle of diffraction of radiation. It was shown that the relative change of the angle of diffraction during temperature change on one degree in the range of measurements error does not depend on the wavelength of the radiation and it is $1.4\cdot10^{-6}$ K^{-1}.

References

[1] Andreeva O and Bandyuk O 2011 Light-Sensitive Media-Composites for Recording Volume Holograms Based on Porous Glass and Polymer *Holograms - Recording Materials and Applications* Dr Izabela Naydenova (Ed.) InTech

[2] Andreeva O, Korzinin Yu and Manukhin B 2013 *Holography - Basic Principles and Contemporary Applications* Dr. Emilia Mihaylova (Ed.) InTech

Estimation of number of resolvable signal levels of photo- and videocameras

P A Cheremkhin[1], V V Krasnov[1], E A Kurbatova[1], V G Rodin[1], S N Starikov[1]
[1]National Research Nuclear University MEPhI (Moscow Engineering Physics Institute), Faculty of Experimental and Theoretical Physics, Department of Laser Physics, Kashirskoe shosse 31, Moscow 115409, Russia

E-mail: holo_mephi@mail.ru

Abstract. Number of signal levels of modern photo- and videocameras is equal to thousands and tens of thousands. However because of temporal and spatial noises of pixels and limitation of linear dynamic range of camera, number of resolvable signal levels is significantly lower. In this paper method of estimation of the number of resolvable signal levels of cameras is proposed and applied to scientific camera.

1. Introduction

Currently photo and video cameras are essential tools in different areas of science and industry. They are used in aerial photography [1-2], digital holography [3-5], information processing [6-7] and etc. Nowadays cameras have ADC with bit depth of 12 to 16 bits. Thus number of signal levels of cameras is equal to thousands and tens of thousands (4096 levels in case of 12 bits, 65536 levels in case 16 bits). However with decrease of pixel size and increase of bit depth of ADC, normalized noise value to one signal value is increasing too in spite of technology improvement. As a result number of resolvable signal levels is significantly lower than maximum number of signal values determined by ADC bit depth. In this paper method of estimation of resolvable signal levels of cameras is proposed and applied to scientific camera.

2. Description of the method of estimation of camera's resolvable signal levels

Number of resolvable signal levels shows the amount of information achievable from each pixel of the camera. It is determined by noise magnitude and value of linear dynamic range of the camera. Higher noise magnitude means wider signal intervals for each resolvable level. To obtain number of resolvable signal levels, six noise and radiometric characteristics of camera's photosensor must be known. These characteristics are:

- noise characteristics of camera's photosensor:
  - light temporal noise,
  - dark temporal noise,
  - light spatial noise,
  - dark spatial noise;
- radiometric characteristic of camera's photosensor:
  - maximum linear signal value,
  - black level offset (BLO, mean value of dark frame).

For estimation of number of camera's resolvable signal levels approximated values of full noise value $\sigma_{\Sigma}(S)$ for each camera's signal value should be determined as follows:

$$\sigma_{\Sigma}(S)=\sqrt{(\sigma_{dt})^2+(\sigma_{lt})^2+(\sigma_{ds})^2+(\sigma_{ls})^2}=\sqrt{(\sigma_{dt})^2+\frac{S}{C}+(DSNU)^2+(PRNU\cdot S)^2} \quad (1)$$

where σ_{dt} - dark temporal noise, σ_{lt} - light temporal noise, σ_{ds} - dark spatial noise, σ_{ls} - light spatial noise, S - signal value with BLO subtracted, C - camera gain (the scaling constant for conversion of digital signal value into the number of electrons), *DSNU* - dark signal non-uniformity, *PRNU* - photo response non-uniformity [8-9]. Light spatial noise is characterized by the *PRNU* value. Dark spatial noise is characterized by the *DSNU* value. Range of signal values *S* is from zero value (BLO is subtracted) to maximum linear signal value.

Number of resolvable signal levels of camera is defined using Eq. (1) as follows. The length of each interval determined by the noise value of the signal *S* considered is equal to double full noise value $\sigma_{\Sigma}(S)$. For calculation of the number of resolvable signal levels an algorithm described below was used:

1. Interval with length equal to noise value at the maximum signal level S_1 of radiometric function linear part $\sigma_{\Sigma}(S_1)$ starts at the signal value S_1. This interval corresponds to the last resolvable signal level.
2. Second resolvable signal level S_2 is found so that $S_2+\sigma_{\Sigma}(S_2)=S_1-\sigma_{\Sigma}(S_1)$ i.e. borders of confidence intervals of the signal values S_1 and S_2 are contiguous. This resolvable signal level S_2 corresponds to interval with length $2\sigma_{\Sigma}(S_2)$. It starts from the signal value $S_1-\sigma_{\Sigma}(S_1)$ which is the confidence interval border of the last resolvable level.
3. The next resolvable signal level S_3 is found so that $S_3+\sigma_{\Sigma}(S_3)=S_2-\sigma_{\Sigma}(S_2)$.

All remaining resolvable signal levels are found similarly until signal level equal to zero value (with subtracted BLO) is reached.

3. Experimental estimation of camera's resolvable signal levels

For experimental estimation of number of camera's resolvable signal levels, scientific camera MegaPlus II ES11000 was used. Its main technical characteristics are given in Table 1 [10].

Table 1. Technical characteristics of camera MegaPlus II ES11000.

Characteristic	Value
Sensor type	CCD
Quantity of pixels	4008 × 2672 (10.7 MP)
Pixel size	9 μm × 9 μm
Output bit depth of ADC	Up to 12 bits
Spectral	Monochromatic
Minimum integration time	140 μs
Maximum integration time	> 5 s

Noise and radiometric characteristics of this camera measured in [11] were used. They are shown in Table 2, where DN is digital numbers of signal, e⁻/DN is the ratio of quantity of emitted electrons in one pixel to one digital number of signal.

Table 2. Noise and radiometric characteristics of camera MegaPlus II ES11000.

Characteristic	Value
Dark temporal noise	2.3 ± 0.4 DN
Camera gain	11.88 ± 0.11 e^-/DN
Dark spatial non-uniformity	0.56 ± 0.04 DN
Photo response non-uniformity	0.0053±0.0008
Black level offset (BLO)	3.4 ± 0.6 DN
Maximum linear signal with BLO subtracted	3916 ± 30 DN
Dark temporal noise	2.3 ± 0.4 DN

Estimated maximum number of camera's resolvable signal levels is equal to 161. This value corresponds to the whole linear dynamic range of the camera. Obtained number of camera's resolvable signal levels is 24 times lower than total number of digital signal levels (3917 levels, see Table 2) defined by the bit depth of the camera ADC for linear part of radiometric function.

It should be noted that obtained value of number of camera's resolvable signal levels is correct for a single shots. The use of averaging over several shots, spatial averaging over pixels groups or compensation of light spatial noise portrait of camera's photosensor [12] leads to noise decreasing and consequently to increasing of number of resolvable signal levels.

4. Conclusion

Method of estimation of resolvable signal levels of photo- and videocameras is proposed. It utilizes iterative approach of consistent estimation of noise of signals values within camera linear dynamic range. The method was applied to scientific camera MegaPlus II ES11000. Estimated maximum number of camera's resolvable signal levels is equal to 161. Proposed method can be used both for digital and analog cameras.

References

[1] Filippi A M and Guneralp I 2013 *Opt. Lett.* **38** 1676-8

[2] Marks D L, Kittle D S, Son H S, Youn S H, Feller S D, Kim J, Brady D J, Golish D R, Vera E M, Gehm M E, Stack R A, Tremblay E J and Ford J 2012 *Opt. Photonics News* **23** 31

[3] Juptner W and Schnars U, 2005 *Digital Holography: Digital Hologram Recording, Numerical Reconstruction, and Related Techniques* (Berlin-Heidelberg: Springer-Verlag)

[4] Mills G A and Yamaguchi I 2005 *Appl. Opt.* **44** 1216-25

[5] Evtikhiev N N, Starikov S N, Cheryomkhin P A, Krasnov V V and Rodin V G 2012 *Proc. SPIE, Optical Modelling and Design II (Brussels)* vol 8429 (SPIE) p 84291M

[6] Chen W, Chen X, Anand A and Javidi B 2013 *JOSA A* **30** 806-12

[7] Gao Y, Tang J, Hong R, Yan S, Dai Q, Zhang N and Chua T-S 2012 IEEE Trans. Image Processing **21** 2269-81

[8] Janesick J 2001 *Scientific Charge-Coupled Devices* (Bellingham-Washington: SPIE Press)

[9] Nakamura J 2006 *Image sensors and signal processing for digital still cameras* (Boca Raton, FL: CRC Press)

[10] Princeton Instruments, http://www.princetoninstruments.com/Uploads/Princeton/Documents/Brochures/MEGAPLUS_camera_brochure_Princeton_Instruments_B0.pdf

[11] Evtikhiev N N, Starikov S N, Cheryomkhin P A and Krasnov V V 2012 *Proc. SPIE, Intelligent Robots and Computer Vision XXIX: Algorithms and Techniques (Burlingame, USA)* vol 8301 (SPIE) p 830113

[12] Evtikhiev N N, Starikov S N, Cheryomkhin P A, Krasnov V V and Rodin V G 2014 *Proc. SPIE, Intelligent Robots and Computer Vision XXXI: Algorithms and Techniques (San Francisco)* vol 9025 (SPIE) p 90250X

Estimation of resolution of dispersive correlator

D Yu Molodtsov[1], V G Rodin[1] and S N Starikov[1]
[1]National Research Nuclear University MEPhI (Moscow Engineering Physics Institute), Kashirskoe shosse 31, Moscow 115409, Russia

E-mail: holo@pico.mephi.ru

Abstract. Estimates of the maximum number of resolvable elements and the relative spectral resolution of dispersive correlators were performed. The obtained relations were used in the synthesis of Fourier holograms. Their validity was confirmed in experiments on recognition of emissive objects in the dispersive correlator setup.

1. Introduction

Optical diffractive correlators are widely used for optical data processing. However, the number of applications for these correlators is usually limited to processing of binary or grayscale images formed in monochromatic light [1]. Unlike correlators for recognizing objects in a monochromatic light, in dispersive correlators [2] recognition signals are formed by the actual radiation of recognized emitting object at its interaction with the spatial filter (optical memory). In these correlators in capacity of spatial filter computer generated Fourier-holograms are often used.

2. Spatial and spectral resolution of dispersive correlator

There are various schemes of dispersive correlators, one of them is presented in figure 1. L_1 is distance from the input plane of the correlator, where emissive recognizable object is located to Fourier-objective. L_2 is distance from objective to output plane, where signal is registered on camera. L is distance from output plane to the plane of spatial frequencies, where synthesized hologram is located. In output plane correlation signal is formed. It takes into account both spatial and spectral parameters of reference and recognized objects. Its location corresponds to hologram first diffraction order.

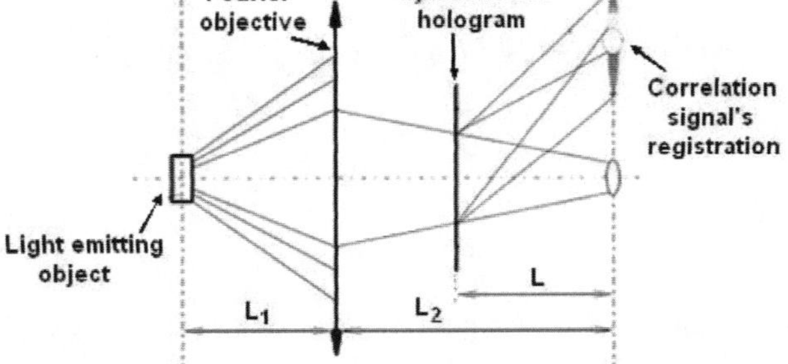

Figure 1. The principal scheme of the dispersive correlator with single Fourier-objective.

The important informational characteristics of the dispersive correlator are: the number of resolvable elements in an image of recognized object (spatial resolution) and resolution on spectrum of radiation. In dispersive correlators part of available pixels is used for recognition of the emission spectrum, so the number of resolvable elements of recognized object image depends on characteristics of its emission spectrum. If synthesized Fourier-hologram is used in correlator, spatial and spectral resolutions are determined by the number of pixels of hologram.

At synthesis of holographic Fourier-filter for dispersive correlator, the generalized spatial pattern of the reference object is recorded on the hologram [3]. This spatial pattern contains such information as the spatial structure and the emission spectrum of the object. This spatial pattern consists of a number of multi-scale copies of reference object's image. The number of copies is defined by the number of components in the emission spectrum of reference object. Scale and location of each copy are inversely proportional to the spectrum component's wavelength

$$\frac{M_{\lambda\,\min}}{M_{\lambda\,\max}} = \frac{P_{\lambda\,\min}}{P_{\lambda\,\max}} = \frac{L_{\lambda\,\min}}{L_{\lambda\,\max}} = \frac{\lambda_{\max}}{\lambda_{\min}}, \tag{1}$$

where $\lambda_{\min}$ and $\lambda_{\max}$ are the minimum and the maximum wavelengths in the spectrum of the reference object, $M_{\lambda\min}$ – the size of the largest copy along the axis of the carrier spatial frequency, $M_{\lambda\max}$ – the size of the smallest copy along the axis of the carrier spatial frequency (the sizes along other coordinate are designated as $P_{\lambda\max}$ and $P_{\lambda\min}$), $L_{\lambda\max}$ – the distance from coordinates origin to the center of the smallest copy, $L_{\lambda\min}$ – the distance from the coordinates origin to the center of the largest copy. In figure 2 the layout of object image's copies on the calculating field of the hologram pulse response is shown. Calculation field has $N\times N$ pixels. Rectangle was used as reference object, its radiation spectrum consists of two wavelengths. $N_{\lambda\min}$ – the distance from the largest copy of the object to the border of the calculating field of the hologram, $N_{\lambda\max}$ – distance from the coordinates origin (the center of zero-order of the hologram) to the left border of the smallest copy of the object.

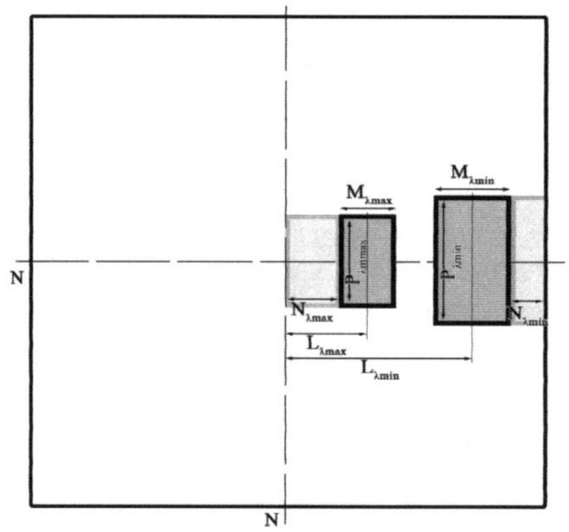

Figure 2. The layout of object image's copies on the calculating field of the hologram pulse response.

If such Fourier hologram is used as a spatial filter, resulting correlation signal is formed in output plane of dispersive correlator. It is the sum of the spectral components of the correlation signals for each wavelength of radiation of recognized object. The values of each spectral component of the correlation signal are determined by the degree of similarity of spatial forms of the reference and recognized objects. Degree of spatial convergence of correlation signal spectral components peaks is determined by degree of similarity of emission spectra of reference and recognized objects.

While estimating the maximum number of resolvable elements it was assumed that the numbers of resolvable elements in the reference and recognized object's images are the same. Correlation signal registration requirements for determining the maximum number of resolvable elements in the image of an object are:

1) zero diffraction order containing the image of an input object should not overlap with correlation signal;
2) the highest diffraction orders should not overlap with correlation signal.

If these requirements are fulfilled, generalized spatial pattern of the reference object on the calculated field of hologram impulse response will require following number of pixels:

$$M_{\text{total}} \le \frac{N}{5}\cdot\left(2-\frac{\lambda_{\min}}{\lambda_{\max}}\right),\ P_{\text{total}} \le \frac{N}{2}. \tag{2}$$

Maximum number of resolvable elements $M\times P$ in image of the reference object is determined by the number of pixels in the smallest copy of reference object's image contained in its generalized spatial pattern. This smallest copy corresponds to the spectral component with a maximum wavelength $\lambda_{\max}$ and has a number of pixels $M_{\lambda\max}\times P_{\lambda\max}$. Therefore, the number of resolvable elements in images of the reference and recognized objects is equal to

$$M \le \frac{N}{5}\cdot\frac{\lambda_{\min}}{\lambda_{\max}},\ P \le \frac{N}{2}\cdot\frac{\lambda_{\min}}{\lambda_{\max}}. \tag{3}$$

The obtained expression is consistent with the analogous expression for diffractive correlator with monochromatic light at $\lambda_{\max} = \lambda_{\min}$.

Estimation of relative spectral resolution of dispersive correlator was obtained from the condition of spatial overlapping of the spectral components of the resulting correlation signal within one resolvable element:

$$\frac{\delta\lambda}{\lambda} = \frac{2}{N}\cdot\frac{\lambda_{\max}}{\lambda_{\min}}. \tag{4}$$

This expression is in agreement with the analogous expression for the spectrum analyzer based on diffraction grating, however it also accounts for the range of the emission spectrum of the reference object $\lambda_{\max} / \lambda_{\min}$. Higher ratio $\lambda_{\max} / \lambda_{\min}$ leads to lower spatial resolution of dispersive correlator.

The obtained expressions were used to estimate informational characteristics of dispersive correlator for recognition of self-luminous objects formed with application of radiation of RGB light-emitting diodes with $\lambda_{\min} = 0.472$ μm and $\lambda_{\max} = 0.645$ μm. The number of pixels of synthesized holograms was limited by phase distortions of used transparent film and was equal 1024×1024 pixels [4]. According to obtained estimates the maximum number of resolvable elements in image of object was equal 149×374 pixels, relative spectral resolution was about 0.003. These values are in agreement with results of experiments on dispersion correlator setup with application of synthesized holograms.

3. Conclusion

Thus, estimates of maximum number of resolvable spatial elements and relative spectral resolution of dispersive correlator were obtained. These characteristics are determined by the number of pixels of synthesized hologram and ratio of maximum to minimum wavelengths in the emission spectrum of recognized object.

References

[1] Cheremkhin P A, Evtikhiev N N, Krasnov V V, Porshneva L A, Starikov S N and Rodin V G 2014 *Proc. SPIE, Optical Modelling and Design III (Brussels)* vol 9131 (SPIE) p 913124
[2] Bykovsky Yu A, Markilov A A, Rodin V G and Starikov S N 1999 *Engineering Physics* **1** 32-5
[3] Rodin V G and Starikov S N 2012 *J. Opt. Technol.* **79** 212-216
[4] Evtikhiev N N, Starikov S N, Cheryomkhin P A, Krasnov V V and Rodin V G 2012 *Proc. SPIE, Optical Modelling and Design II (Brussels)* vol 8429 (SPIE) p 84291M

Numerical reconstruction of wave field spatial distributions at the output and input planes of nonlinear medium with use of digital holography

S S Nalegaev[1], N V Petrov[1], V G Bespalov[1]
[1]Saint Petersburg National Research University of Information Technologies, Mechanics and Optics, Kronverkskiy pr. 49, St. Petersburg 197101, Russia

E-mail: s.nalegaev@gmail.com

Abstract. A numerical reconstruction of spatial distributions of optical radiation propagating through a volume of nonlinear medium at input and output planes of the medium was demonstrated using a scheme of digital holography. A nonlinear Schrodinger equation with Fourier Split-Step method was used as a tool to propagate wavefront in the volume of the medium. Time dependence of the refractive index change was not taken into account.

1. Introduction

It is known that optical systems have a limited field of view due to a finite value of the refractive index of lenses and a finite size of apertures. That is why, in the linear case, high spatial frequencies of the input radiation miss the detector. In the nonlinear case, these spatial frequencies after their nonlinear wave mixing with other components of the input radiation can create daughter waves propagating at a different angle, which allows these spatial frequencies to get to the detector [1]. The nonlinear processes are also accompanied by spectral broadening or spectral supercontinuum generation [2–4]. It is reported about a possibility to develop new methods to achieve the optical super-resolution based on the effect of the nonlinear wave mixing [5], because in the nonlinear case the recorded wavefront keeps the information about non-detected high spatial frequencies as well as evanescent waves of the input radiation. From this perspective, a development of a numerical model for imaging through a volume of nonlinear media has some point of interest.

2. Numerical reconstruction scheme

Wavefront distortions caused by nonlinearity depend on the parameters of the input radiation and the nonlinear medium [6]. In this paper, using digital holography method a numerical model was developed that allows to restore wave field at the output plane of the nonlinear medium from the known digital hologram and to get the original wave field at the input plane by back-propagation of previously calculated wave field. It is possible if a nonlinear response of the medium is known [6–8]. It was assumed that monochromatic laser source with continuous wave mode was used, therefore, a time dependence of the intensity induced refractive index change was not taken into account.

The wavefront reconstruction at the output plane of the nonlinear medium from the known digital hologram was performed in accordance with the method described in [9]. In the scheme of our numerical experiment two 4f-systems were used to transfer a wavefront from the test object O to the input plane of the nonlinear medium and from the output plane to the CCD photodetector (Figure 1).

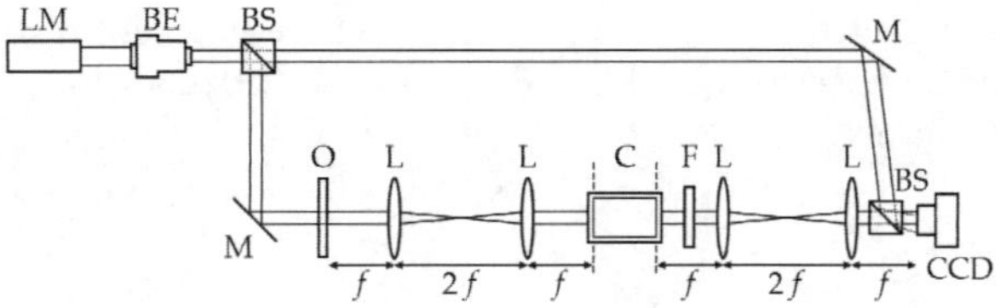

Figure 1. An experimental setup scheme, illustrating the numerical experiment for wave field reconstruction in the volume of nonlinear media using an off-axis digital holography method: LM – laser, BE – beam expander, M – mirror, BS – beam splitter, O – test object, L – lenses, C – cuvette with tea in water solution (nonlinear medium), F – optical filter, CCD – photodetector.

To numerically restore a wave field at the input plane of the nonlinear medium using nonlinear digital holography, one needs to know a complex field of the radiation at some point. This complex field can be numerically propagated forward or backward [6] in the volume of the medium if the mathematical model of the propagating process is fairly accurate. Within a paraxial approximation the dynamics of the wave field of slowly varying wave function U is well described with nonlinear Schrodinger equation [10,11]:

$$\frac{\partial U}{\partial l} = \left[\mathrm{i}\frac{1}{2k_0}\nabla^2{}_\perp + \mathrm{i}k_0 \Delta n(I) \right] U \equiv \left[D + N(U) \right] \cdot U \tag{1}$$

U – complex wave field; l – propagation direction; i – imaginary unit; $k_0 = n_0(2\pi/\lambda)$ – wave number; n_0 – linear refraction index; λ – wavelength; $\nabla^2{}_\perp$ – nabla; $\Delta n(I) = n_2 I$ – nonlinear refractive index change; n_2 – nonlinear refraction index; $I = n_0 \cdot E_0 \cdot c \cdot |U|^2/2$ – intensity; D and N – linear and nonlinear operators (the angular spectrum method [12,13] was used to calculate diffraction part D).

3. Results

Numerical simulation was performed with following parameters: λ = 633 nm; object size 2×2 mm (512×512 pixels); thickness of the nonlinear medium 7.5 mm; nonlinear step Δl = 50 um along the l direction; optical power 1 W; linear n_0 = 1.33 and nonlinear $n_2 = -1.8\times10^{-11}$ m^2/W refraction indexes.

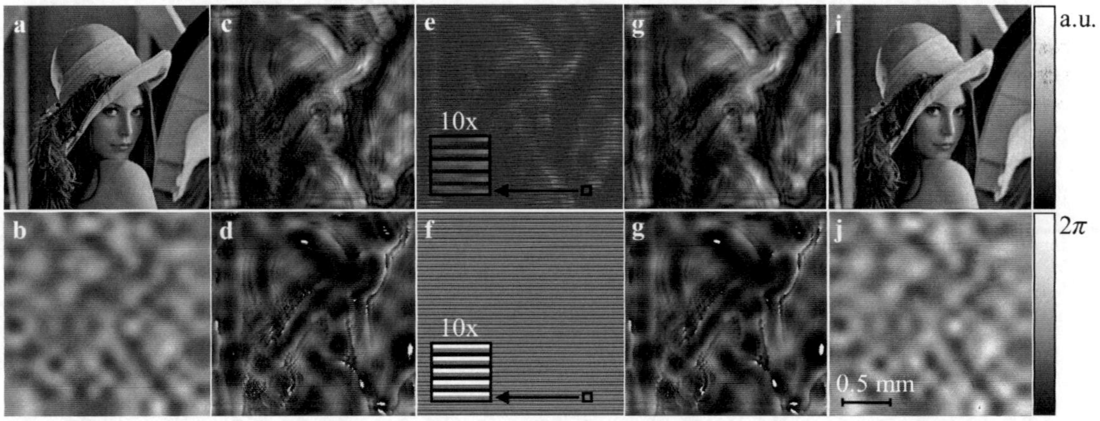

Figure 2. Numerical retrieval amplitude and phase distributions of the wave field at the output (g,h) and input (i,j) planes of the nonlinear medium from the known digital hologram (e,f) of intersecting object (c,d) and reference beams (selections on pictures are enlarged by 10 times) with initial wave field distribution at the input plane (a,b).

Specified n_0 and n_2 parameters of the nonlinear medium correspond to the tea in water solution at the wavelength $\lambda \approx 633$ nm [14], the nature of its nonlinearity has a thermal origin. Results of the numerical experiment are presented above (see Figure 2). The wave field distribution at the output plane of the nonlinear medium is determined by the balance of effects of diffraction and nonlinear defocusing that manifest themselves on the heterogeneity of initial distributions of amplitude (a) and phase (b). The numerical recovery algorithm can work with arbitrarily specified initial distributions.

4. Conclusion

A reconstruction of the wave field at the output and input planes of the nonlinear medium from the recorded digital hologram was numerically demonstrated. The method [9] was used to reconstruct wave field from the digital hologram. A wavefront back-propagation in the volume of the nonlinear medium was performed by the Fourier Split-Step method as a numerical solution of the nonlinear Schrodinger equation. Time dependence of the refractive index change was not taken into account.

The use of nonlinear methods in optics is of interest due to the fact that restoring nonlinear dynamics of the field and retrieving the missing spatial frequencies provides an opportunity to develop new methods for the optical super-resolution, including methods for optical sub-wavelength microscopy and lithography, microscopy with resolution exceeding diffraction limit. In this case, the effect of super-resolution results from a basic knowledge about the function of nonlinear propagation.

Acknowledgments

Sergey S. Nalegaev and Nikolay V. Petrov acknowledge financial support from Russian Ministry of Education and Science project within the state mission for institutions of higher education (Agreement No 2014/190). Victor G. Bespalov acknowledges financial support from Government of Russian Federation, Grant No 074-U01.

References

[1] Barsi C and Fleischer W J 2010 *Conf. on Lasers and Electro-Optics: General Aspects of Nonlinear Optics 2010* (*San Jose, CA, 16-21 May 2010*) OSA Technical Digest (CD) p CMCC4
[2] Alfano R R 2005 *The Supercontinuum Laser Source: Fundamentals with Updated References* (New York: Springer-Verlag)
[3] Nalegaev S S, Putilin S E, Bespalov V G 2012 Particularities of femtosecond spectral supercontinuum generation in crystal media with $X^{(2)}$-nonlinearity *Scientific and technical journal of information technologies, mechanics and optics* **81** 29–32
[4] Nalegaev S S, Putilin S E, Bespalov V G 2013 Particularities of femtosecond spectral supercontinuum generation in anisotropic crystal media with quadratic nonlinearity *Saratov Fall Meeting 2012: Optical Technologies in Biophysics and Medicine XIV; and Laser Physics and Photonics XIV* ed. Tuchin V V, Genina E A, Derbov V L, Meglinski I V (Bellingham: Proceedings of SPIE vol 8699) p 869914-1–869914-6
[5] Puida M and Ivanauskas F 2005 *Liet. matem. rink.* **45** 504–8
[6] Barsi C, Wan W and Fleischer J W 2009 *Nature Photonics* **3** 211–5
[7] Tsang M, Psaltis D and Omenetto F G 2003 *Opt. Lett.* **28** 1873–75
[8] Nalegaev S S, Petrov N V, Bespalov V G 2012 Special features of iteration methods for phase problem in optics *Scientific and technical journal of information technologies, mechanics and optics* **82** 30–5
[9] Liebling M, Blu T, Unser M 2004 *J. Opt. Soc. Am* **A21** 367–77
[10] Whitham G B 1974 *Linear and Nonlinear Waves* (New York: Wiley)
[11] Agrawal G P 2013 *Nonlinear Fiber Optics* (New York: Academic Press)
[12] Goodman J W 2005 *Introduction to Fourier Optics* (Englewood, CO: Roberts & Company)
[13] Poon T C, Banerjee P P 2001 *Contemporary Optical Image Processing with MATLAB* (Elsevier Science & Technology)
[14] Oliveira L C, Zilio S C 1994 *Appl. Phys. Lett.* **65** 2121-23

Journal of Physics: Conference Series **536** (2014) 012026 doi:10.1088/1742-6596/536/1/012026

Numerical study of density fluctuations of suspended particles in a volume of optical medium

T Nikolaeva[1], **N Petrov**[1]
[1]Saint Petersburg National Research University of Information Technology, Mechanics and Optics, Kronverkskiy pr, 49, St. Petersburg 197101, Russia

E-mail: nikolaevatatiana.yu@gmail.com

Abstract. This paper is devoted to the numerical study of density fluctuations of scattering particles suspended in a volume of the optical medium. Two spatial distributions were considered: random distribution and normal distribution. As a result of statistical processing of the focused images of scattering particles normalized distributions of density in the image plane, which characterize the fluctuations in the volume of the optical medium were obtained.

1. Introduction

The necessity for statistical accounting of particles in the volume of the medium arises in many scientific and practical problems, for example, in problems of atmospheric optics in the study of gas dynamics of particle fluxes [1], in oceanography and biology in the study of particles or small organisms suspended in the fluid [2] and in the study of transparent media. [3] There are several methods by which it is possible to quantitatively and qualitatively examine the distribution of particles in the optical medium. Methods of processing of images of particles obtained by coherent illumination currently became more widespread. Depending on the concentration of the particles contained in the volume, scattered laser radiation forms two distinct structures, which in turn leads to two possible approaches to image processing: processing of images of individual particles carried out by means of PIV (particle image velocimetry) [4] or processing of speckle patterns using DSP or ESP-methods (digital/electronic speckle photography) [5]. Formation of a speckle pattern leads to serious limitations: even a relatively small movement of the particles outside the illuminated plane can lead to a serious change in the speckle pattern between exposures.

The aim of this work was to study density fluctuations of scattering particles suspended in the volume of optical medium by numerical simulation. Using statistical accounting of particle images, normalized density distributions of particles in the image plane that determine their fluctuations in the volume of the optical medium were obtained.

2. Statistical study of scattering particles suspended in the volume of optical medium

In this paper the propagation of electromagnetic radiation through the volume of the optical medium with different statistical distributions of particles in it is simulated. Figure 1 shows a schematic illustration of an optical system that forms the image of a volume of optical medium with particles. The electromagnetic radiation passing through the volume of the medium which is previously broken into segments (a, b, c - in Figure 1) is diffracted by the particles. The concentration and the size of particles may vary for different experimental implementations. Such spatial distributions of particles as random (equiprobable in each segment) distribution and normal distribution, which is realized in

Journal of Physics: Conference Series **536** (2014) 012026 doi:10.1088/1742-6596/536/1/012026

the numerical model using the Box-Muller transform (random variables are normally distributed with respect to the optical axis) are considered in this paper. The method of distribution of the angular spectrum of plane waves was used to calculate the electromagnetic field propagation through the volume of the optical medium and optical system.

Image of particles in a certain volume segment (segment b), is transferred to the recording plane by a lens. Recorded image of the particles (3 – Figure 1) is the result of a coherent superposition of the particles located in different volume segments. In the numerical experiment, it was possible to obtain focused images of different planes of the volume of the optical medium by changing the distance on which the propagation of the wave field from the plane of volume segment to the lens plane is calculated. In this case, the images of the neighboring planes take on the form of defocused spots.

Figure 2 shows the results of the simulation of the focused images of particles with different concentrations of N in one transverse volume segment for random and normal distributions of particles. Simulated volume of the medium 0,768×1,536×1,536 mm during the numerical experiment was divided into five segments, grid size Δx = 6 μm, wavelength λ = 632,8 nm, focal length of the imaging lens f = 29,1mm.

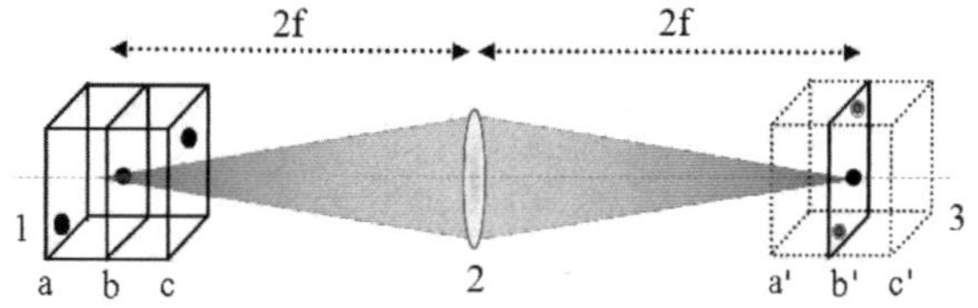

Figure 1. The intensity distribution of a wave field diffracted by three particles in the volume of optical medium is recorded in the segment b': 1 – volume of the medium divided into segments (a, b, c), 2 – lens, 3 – focused images of particles a', b', c'.

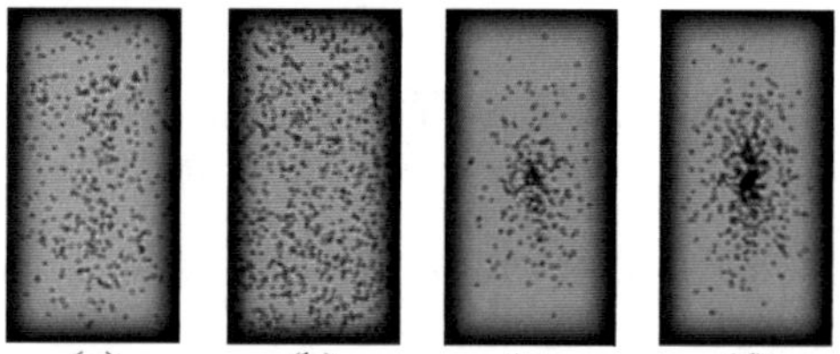

Figure 2. Focused particles images, (a)-(b) – random distribution, (c)-(d) – normal distribution, (a), (c) – N=50, (b), (d) – N=100.

In this paper the statistical analysis of the focused particle images is carried out. For automatic image recognition of focused particles the following algorithm was developed and implemented in the code: in the first stage, approximate regions with focused particle images are obtained; in the second stage, the program finds double brightness jumps from light to dark by scanning the selected image fragment and determines the coordinates of the particles; in the final stage, on the basis of the data obtained, the radial and density distributions of particles in the image plane are constructed, which determine fluctuations of density in the volume of the optical medium (Figure 3).

If the density of particles in one segment of the medium volume is low, then all focused particles can be detected. With increasing number of particles per segment of the volume, the effect of coherent superposition of the particle images of the adjacent planes begins to appear strongly, which leads to a partial loss of information about the particles in the test segment of the volume. After reaching the threshold value under certain specified parameters, we obtain not the individual particle images but the speckle pattern in the image plane. In this case, as in the case of studying the particles which are transparent to the probing radiation, but introduce the phase retardance (phase objects), the detection (contouring) of the particles using the method of analysis of singles image becomes problematic. As shown in Figure 3, random distribution of detected particles (a,b) at N = 150 is well approximated by Poisson distribution compared to normal distribution at the same concentration (c,d). On the normalized distribution of the mass fraction of particles for normal distribution (Figure 3-d), the additional peaks of the distribution function appear, which can be explained by the formation of clusters of particles near the optical axis as a consequence of the specifics of used distribution law.

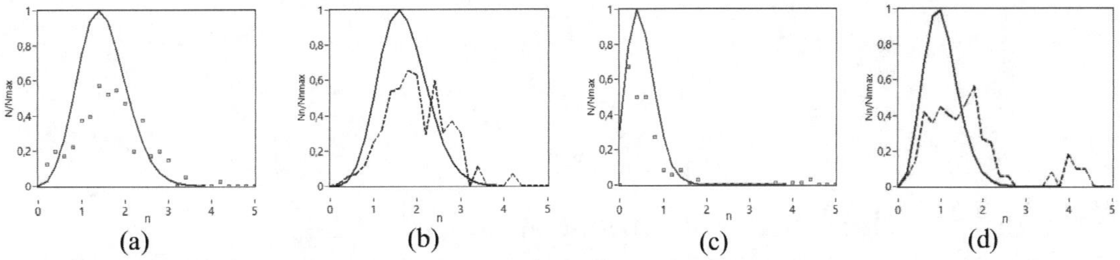

Figure 3. (a), (b) – random distribution, (c), (d) – normal distribution (N=150); (a), (c) - calculated according to Poisson statistics (solid curve) and experimental (indicated by dots) normalized density distribution of the particles in the image plane; (b), (d) - calculated according to Poisson statistics (solid curve) and experimental (dashed curve) normalized distribution of the mass fraction of particles.

3. Conclusion

In this work, the density fluctuations of the scattering particles in the volume of an optical medium were investigated. As a result of applying method of statistical accounting of particle images, based on the contouring of particles and thresholding technique, normalized density distributions of particles in the image plane that determine their fluctuations in the volume of the optical medium were obtained. At low particle densities in the volume of the medium, recognition of images of particles in the particular segment is lossless. With the increase of the amount of particles per segment of the volume, the effect of coherent superposition of the images of particles in adjacent planes becomes apparent. Reaching the threshold concentration of particles per segment results in the inability to obtain complete information about the statistics of the particles in the test segment of the volume.

Acknowledgments

Nikolay Petrov acknowledges financial support from Russian Ministry of Education and Science project within the state mission for institutions of higher education (Agreement No 2014/190)

References

[1] Voronetsky A V, Mikhailov V N, Petrov N V, Staselko D I 2012 *J. Opt. Technol.* **79** 18–24.
[2] Dyomin V V, Olshukov A S 2012 *J. Opt. Technol.* **79** 344-347
[3] Bazulev N et al. 2003 *Laser Phys.* **13** 786–795
[4] Adrian R J, Westerweel J 2011. *Particle Image Velocimetry* (Cambridge: Cambridge University Press)
[5] Sjödahl M 1994 *Appl. Opt.* **33** 6667–6673

A method for the coherence measurement of the supercontinuum source using Michelson interferometer

V A Semenova[1], A V Tsypkin[1], S E Putilin[1] and V G Bespalov[1]
[1]Saint Petersburg National Research University of Information Technologies, Mechanics and Optics, 49 Kronverkskiy pr., St. Petersburg 197101, Russia

E-mail: barb.semenova@gmail.com

Abstract. Coherent properties of supercontinuum sources are highly significant for various applications, including low-coherence interferometry and optical frequency metrology. We propose a fast method for the spatial and temporal self-coherence of the SC measurement using Michelson interferometer without a mirror movement. Furthermore, we present self-coherence measurements of the supercontinuum, generated in microstructured fiber at 780 nm.

1. Introduction

Supercontinuum (SC) generated in microstructured fiber is a unique source of broadband light, which can span for more than an octave, and now it is an indispensable part of various applications [1]. But not only the SC ultrabroad bandwidth is an important parameter to be considered for any given application, the spectral and temporal coherence characteristics are also of primary interest [2]. Therefore, the coherence properties of SC have been studied intensively for the last decade both theoretically and experimentally [1-5].

All experiments for the coherence properties of SC measurements can be divided into two groups depending on whether the 'same' pulse spectral coherence or "pulse-to-pulse" mutual coherence is measured. In the first case the interference is provided with the pair of the same SC pulses, separated into the reference and sample pulses, and it is the self-coherence of the supercontinuum. The resulting interference pattern depends on time-delay interference that is averaged over multiple pairs [1]. The mutual coherence results in interference between independently generated supercontinua pulses.

A number of methods to measure self-coherence properties of the SC have been demonstrated for both spatial and temporal domains [1,5]. Spatial spectral coherence is usually measured with the Young-type double-split experiments [5].The conventional setup for the spectral-temporal coherence measurement includes diffraction-grating-based interferometer or the Michelson-type interferometer [1]. When the Michelson-type interferometer is used, the reference and object mirrors are set parallel and the delay between pulses is obtained by the object mirror movement. The interference between pulses results in the circular interference pattern on the interferometer exit. An intensity distribution versus path delay is obtained by measuring the intensity in the center of interference pattern at different positions of the object mirror.

This paper presents another method for the temporal self-coherence of supercontinuum measurement using Michelson interferometer. Note that the spectral spatial coherence of the SC can also be measured by this method.

Journal of Physics: Conference Series **536** (2014) 012027 doi:10.1088/1742-6596/536/1/012027

2. Method description

The temporal self-coherence of a femtosecond supercontinuum was measured with a compact Michelson setup. But in contrast there was a small angle α between object and reference mirrors [6]. The interference between reference and objective pulses results in isoclinic fringes in the image plane. Since the interferometer mirrors are angled, there is a temporal delay between different points of the pulses' wave fronts. There is a zero delay for the central fringe in the interference pattern and the other fringes' delay Δ is given by

$$\Delta = \pm m\lambda , \tag{1}$$

where m – diffraction order of the fringe, λ – radiation wavelength. Visibility of fringes decreases from the center to the edges of the interference pattern accordingly to the increase of the delay between the pulses. Visibility of the fringes is given by

$$V = \frac{I_{\max} - I_{\min}}{I_{\max} + I_{\min}}, \tag{2}$$

where I_{max} and I_{min} is the maximum and minimal intensity of the fringe, respectively. Thus, one can measure the module of the complex degree of coherence without any mirror movement. Spectral coherence of the SC can also be obtained from the resulting interference pattern.

3. Experimental setup

Figure 1 shows the optical schematic of the experimental system configured in this investigation. A Ti:Sapphire femtosecond laser of a 780 nm center wavelength is used as the light source to generate pulses of ~ 80 fs duration. After going through the dispersion compensator, duration of the pulses was 10-20 fs. For SC generation through nonlinear processes, a microstructured crystal fiber MS-38 is used. The 30 cm length fiber is consisted of a fused silica core, embedded in a fused silica cladding of air-filled cylinders parallel to the core in a hexagonal pattern. The SC spectrum was measured with ASP100 spectrometer with detection range 190-1100 nm. To measure the coherence properties of the SC, the Michelson interferometer with angled mirrors was used. The intensity distributions on the interferometer exit were registered using digital commercial camera Canon 1000D without an objective. All the photos were taken in RAW format to achieve a linear intensity distribution.

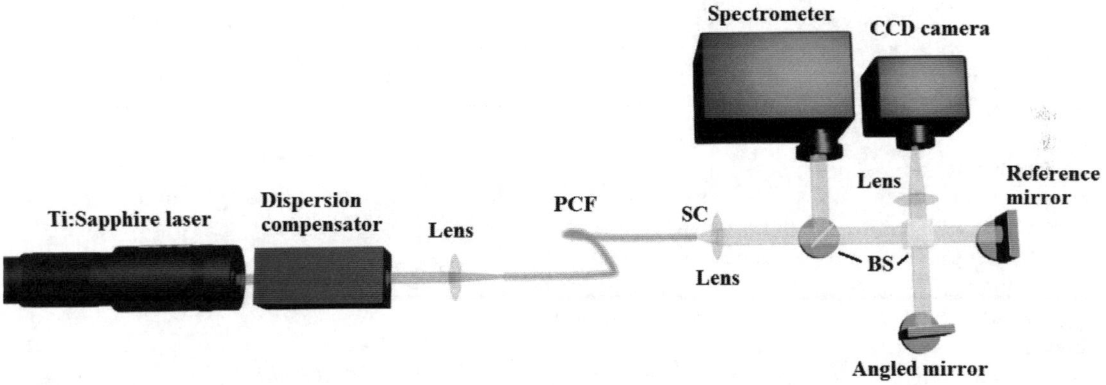

Figure 1. Experimental setup for spectral-temporal coherence of the supercontinuum generation measurements. PCF – microstructured fiber; SC – supercontinuum pulse; BS – beam splitter.

Figure 2 shows a recorded spectrum of the SC on the microstructure fiber exit.

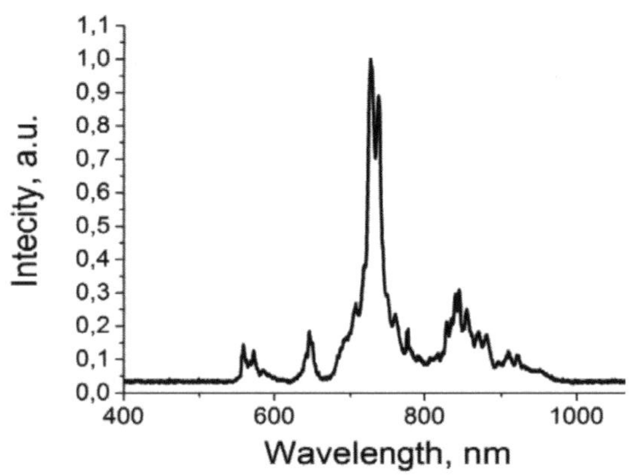

Figure 2. A recorded spectrum of the SC at a pump wavelength of 780 nm. The SC spectrum covers the range from 550 to 900 nm, spanning from green to near infrared.

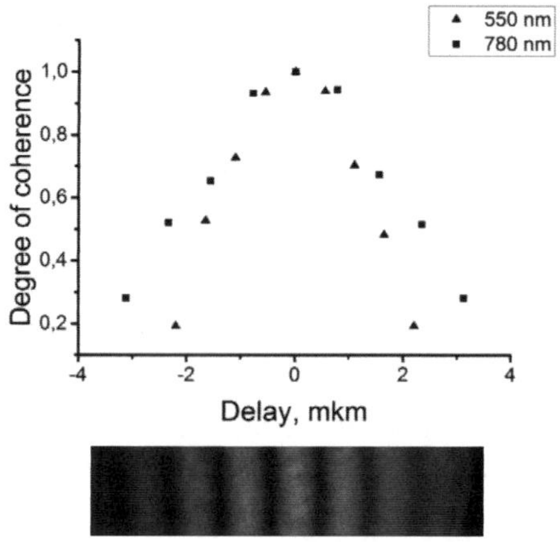

Figure 3. A - Calculated module of complex degree of coherence versus path delay in mkm. B – Interference pattern on the interferometer exit.

4. Data processing

Solid photosensors in CCD matrix present RGGB-structure of Bayer colored filters array. Thus, the incoming light is divided into three channels: red, green and blue. To extract the linear data from the RAW snapshots, the open-source program converter DCRAW was used. Linear file is a grayscale image, which repeats RGGB-structure of Bayer colored filters array. From the grayscale image data three images were allocated in LabVIEW, corresponding to the different color filters. To minimize the error caused by wavelength dependence of the matrix transmittance, these three images were summarized with calculated weight coefficients. To obtain the module of the complex degree of coherence, the visibility of the fringes across the image were calculated using equation (2). The result is shown in figure 3A.

5. Conclusion

Method of estimation of the spatial and temporal self-coherence of the supercontinuum source is proposed. It utilizes the Michelson interferometer with angled mirrors. The method was used to measure spectral-temporal coherence of the femtosecond supercontiuum source. Estimated temporal coherence length is about 2.3 mkm that is in a good agreement with theoretical value of 2 mkm.

References

[1] Alfano R R 2006 *The Supercontinuum Laser Source: Fundamentals with Updated References* (New York: Springer)

[2] Genty G, Surakka M, Turunen J and Friberg A T 2011 *J. Opt. Soc. Am.* B **28** (9) 2301-09

[3] Korhonen M, Turunen J, Friberg A T and Genty G 2013 *J. Opt. Soc. Am.* B **30** (1) 21-6

[4] Chick B, Chon J and Gu M 2009 *Opt. Express* **17** (22) 20140-48

[5] Gu X, Kimmel M, Shreenath A P, Trebino R, Dudley J M, Coen S and Windeler R S 2003 *Opt. Express* **11** (21) 2697-2703

[6] Connes J 1961 *Rev. Opt.* **40** 45-78

Experimental comparison of phase retrieval methods which use intensity distribution at different planes

I A Shevkunov[1], N V Petrov[1]

[1] Saint Petersburg National Research University of information Technologies, Mechanics and Optics, Kronverkskij pr, 49, Saint-Petersburg 197101, Russia

E-mail: Shevkunov_Igor@mail.ru

Abstract. Performance of the three phase retrieval methods that use spatial intensity distributions was investigated in dealing with a task of reconstruction of the amplitude characteristic of the test object. These methods differ both by mathematical models and order of iteration execution. The single-beam multiple-intensity reconstruction method showed the best efficiency in terms of quality of reconstruction and time consumption.

1. Introduction

Full wavefront reconstruction from intensity distributions is receiving a lot of attention lately. This is due to the fact that the methods to the solution of this problem, though usually more complex data processing algorithm, but impose significantly lower requirements for the experimental setup, as opposed to the more traditional holographic approaches. The first phase retrieval iterative algorithm was proposed by Gerchberg and Saxton [1]. They used spatial amplitude distribution in the plane of the object and in the Fourier plane. Its demerits as stagnating and sensitivities to the introduction of errors and noise have been overcome by use of additional datasets in iterative procedure [2]. Most simple example of obtaining such datasets is the registration of intensity distributions in different planes [3-5]. These methods differ by mathematical models and how to use the intensity distributions in the iterative algorithm. The purpose of this work is experimental comparison of efficiency of these methods for amplitude object.

2. Description of the methods

1) Fresnel Iterative Method (FRIM) [3] uses the 2D Fresnel transform in mathematical model:

$$U(x,y,l)=\frac{\exp(ikl)}{i\lambda l}\iint_{-\infty}^{+\infty}\exp\left(\frac{ik}{2l}\left[(x-x')^2+(y-y')^2\right]\right)u(x',y')dx'dy' \quad (1)$$

Here $U(x, y, l)$ is a complex-valued wave field at the distance l; x, y are coordinates in the imagine plane; $u(x', y')$ is the complex characteristic of the object; x', y' are coordinates in the object plane; λ is the wavelength, $k = 2\pi/\lambda$ is the wave vector. The order of the use of measured intensity distributions in the FRIM is following: $I_1 \rightarrow O \rightarrow I_2 \rightarrow O \rightarrow I_3 \rightarrow O \rightarrow \ldots \rightarrow O \rightarrow I_n \rightarrow O \rightarrow I_1 \ldots$. Here $I_1, I_2, I_3, \ldots, I_n$ are spatial intensity distributions recorded at different planes on various distances from the object; arrows denotes wavefront propagation at free space; $O=v(x',y')$ is the object approximation function at the object plane. When convergence will take place this function coincides with $u(x',y')$.

2) The single-beam multiple-intensity reconstruction (SBMIR) method [4] uses the angular spectrum approach in solving the Rayleigh–Sommerfeld equation, which simulates wave propagation between the adjacent measurement planes. The first iteration includes one complete set of sequential propagations over the n planes in the forward direction, Before the 2nd iteration is carried out, the wavefront at the last measurement plane is backpropagated directly to the 1st plane, i.e. $I_1 \to I_2 \to \ldots \to I_n \to I_1 \ldots$. When convergence will be reconstructed wavefront is propagated to the object plane to obtain object function $u(x',y')$.

3) Forward-Backward (FB) SBMIR method [5] differ from conventional SBMIR method by the iterative calculation procedure: $I_1 \to I_2 \to I_3 \to \ldots \to I_{n-1} \to I_n \to I_{n-1} \to \ldots \to I_3 \to I_2 \to I_1 \ldots$.

3. The experiment

Figure 1 shows the object plane and the intensity measurement planes. As radiation source we use a single-mode laser Lasos RLD F-638-50-pvc, λ = 634.9 nm. The test object logo "*ITMO*" was made on high-resolution film (depicted at figure 2). The size of the object 0.5×0.6 mm, letters height 0.2 mm. To register intensity distributions the CMOS matrix "VEI - 830" with resolution 2048×1536 and pixel size Δx = 2.8μm was used. For precise displacements of the sensor, a motorized micrometer stage Standa 8MT175 was used. Three intensity measurement were used in the experiment for each method. Distance l_0 = 17.8 mm, and distances between measurement planes were chosen in according to the criterion that gives the best quality of reconstruction as defined in [6]. Therefore the distance between the first and the second planes was 0.81 mm, the distance between second and third planes was 0.90 mm.

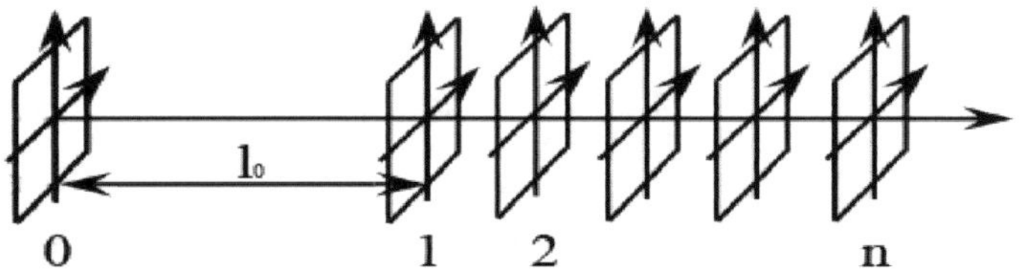

Figure 1. Diagram for measurements datasets from spatial intensities distributions.

Figure 2. Test object used in the experiment.

4. Obtained results

Figure 3 shows the reconstructed images of the object "ITMO" by three different methods: FRIM (a), SBMIR (b), FB-SBMIR (c). Due to the features of FRIM iterative procedure, associated with alternating calculation of wavefront propagation from registration planes to the object plane and back, diffraction artifacts on the boarders of the reconstructed image (figure 3a) appears. Comparing the results on the figure 3, it can be seen that the reconstructed images are almost identical for all three methods. This is well illustrated in Figure 4, which shows the cross section of the enlarged section letters "IT", indicated by the red line on figure 3c. As can see from figure 4, SBMIR and FB-SBMIR showed very similar results, but the method FB-SBMIR more time consuming.

a b c

Figure 3. Reconstructed images: a – FRIM, b – SBMIR, c – FB-SBMIR

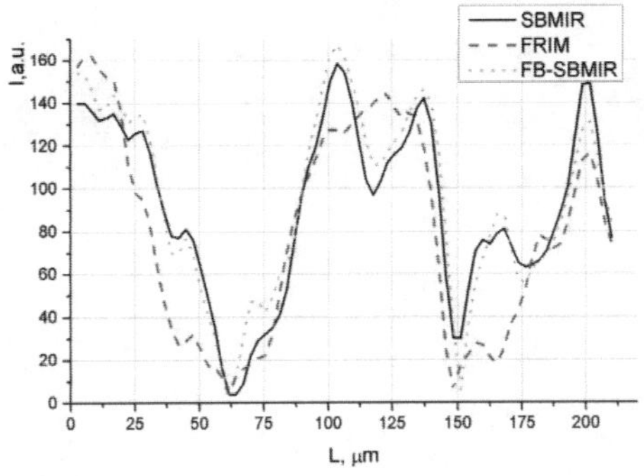

Figure 4. Transverse object amplitude distribution, reconstructed by different methods: FRIM(solid black line), SBMIR(dashed red line) and FB-SBMIR(dotted blue line).

5. Conclusion

The comparison of the effectiveness of three iterative phase retrieval methods in relation the reconstruction of complex characteristics of the test object was made. On the reconstructed image by FRIM methods the diffraction artifacts appears. The reason is iteratively repeating process of the wavefront propagation between object and registration planes. A similar process take place inside laser cavity, which leads to the formation of high-order transverse electromagnetic modes described by Hermite-Gaussian functions. The results, obtained in the experiment by SBMIR and FB-SBMIR methods are similar. But FB-SBMIR method was developed for dynamic wavefront sensor setup with stack of beamsplitters providing instant registration if all intensity distributions. In this case the criterion of the registration of datasets defined in [6] that gives the best quality of reconstruction is difficult to apply in practice. Therefore, we can conclude that the best performance demonstrates the SBMIR method.

Acknowledgments

Authors acknowledge financial support from Russian Ministry of Education and Science project within the state mission for institutions of higher education (Agreement No 2014/190).

References

[1] Gerchberg R W and Saxton W O 1972 *Optic* **35** 237
[2] Petrov N V *et al* 2011 *in Frontiers in Optics 2011/Laser Science XXVII* JTuA9
[3] Petrov N V *et al* 2011 *Proc. SPIE* **7907** 790718
[4] Almoro P *et al* 2006 *Appl. Opt.* **45** 8596
[5] Almoro P *et al* 2009 *Appl. Opt.* **48** 1485
[6] Petrov N V *et al* 2012 *Proc. SPIE* **8413** 84131T

AUTHOR INDEX